Contents at a Glance

TechTV's Guide to the Golf Revolution: How Technology is Driving the Game

International Standard Book Number: 0735714061

Library of Congress Catalog Card Number: 2003111379

Printed in the United States of America

First printing: December, 2003

08 07 06 05 04 03   7 6 5 4 3 2 1

Interpretation of the printing code: The rightmost double-digit number is the year of the book's printing; the rightmost single-digit number is the number of the book's printing. For example, the printing code **03-1** shows that the first printing of the book occurred in **2003**.

Trademarks

Warning and Disclaimer

Bulk Purchases/Corporate Sales

The publisher offers discounts on this book when ordered in quantity for bulk purchases and special sales. For sales within the U.S., please contact: Corporate and Government Sales at (800)382-3419 or **corpsales@pearsontechgroup.com**. Outside of the U.S., please contact: International Sales at (317)581-3793 or **international@pearsontechgroup.com**.

650 Townsend Street

San Francisco, California 94103

Publisher

Nancy Ruenzel

Associate Publisher

Stephanie Wall

TechTV, Vice President, Strategic Development

Glenn Farrell

Production Manager

Gina Kanouse

TechTV Press Project Manager

Sasha Zullo

Acquisitions Editor

Wendy Sharp

Project Editor

Michael Thurston

Copy Editor

Keith Cline

Senior Indexer

Cheryl Lenser

Composition

Gloria Schurick

Manufacturing Coordinator

Dan Uhrig

Interior Designer

Alan Clements

Cover Designer

Aren Howell

Marketing

Scott Cowlin
Tammy Detrich
Hannah Onstad Latham

Publicity Manager

Susan Nixon

Table of Contents

About the Author

Andy Brumer has written for virtually all of golf's major magazines, including *Golf Digest*, *Golf Magazine*, *Golf Illustrated*, *Sports Illustrated Golf Plus*, *Travel and Leisure Golf*, *Links Magazine*, *Golfweek*, *The Golfer*, *Golf for Women*, *Golf Tips*, and others. His golf writing has also appeared in other magazines and newspapers, including *The Los Angeles Times*, *The Wall Street Journal*, *Westways Magazine*, *Sport Magazine*, *Luxury Living Magazine*, *Stratos Magazine*, and others. He has been editor of *Golf Tips* and *Petersen's Golfing*.

Andy also writes extensively on fine art and literature and publishes book reviews in *The New York Times Book Review* and other publications. He is the author of *Turtle*, a national award-winning book of poetry.

Andy currently holds a four handicap. He played Division One golf at Rutgers University before transferring to the University of Wisconsin, where he earned his B.A. degree in English Literature. He also holds an M.A. in Creative Arts from San Francisco State University.

He lives in Alhambra, California, near Los Angeles.

Photo by Ray Ferro

Acknowledgments

I would like to thank many people for their help in preparing and writing this book. First, I'd like to express my gratitude to my agent, John Monteleone. Next, TechTV and Peachpit Press, for their decision to publish a book about golf. Wendy Sharp, the book's editor, not only exercised her remarkable talent in fine-tuning the text, texture, and format of this book, but she showed formidable patience in dealing with me under the pressure of a tight deadline. Sasha Zullo of TechTV provided invaluable research and a willingness to help in any way asked.

This book could not have happened without the generosity of everyone who provided its many sidebars. These busy men and women, all experts in one aspect of the game of golf or another, took time out from their busy schedules to share their knowledge, insights, and passion for the game with me. I'd be remiss if I didn't single out a couple who went the extra mile to make sure that the manuscript was correct and sufficiently comprehensive. First, Benoit Vincent of TaylorMade-adidas golf, whose grasp of golf equipment technology is equaled only by his capacity and creativity in developing wonderfully performing golf clubs. Also, Kit Mungo, an extraordinary club-repairman, who helped me understand technology in a way that could be clearly communicated to the golfer/reader. Tim Moraghan, of the USGA, kindly directed me to many of the game's top experts in golf agronomy, construction, and maintenance; and golf course architect Steve Timm did the same in the field of golf course architecture.

Lastingly, more than lastly, I'd like to thank a group of people—ironically, only two of whom are golfers—who gave me the kind of emotional support and love that made it possible for me to keep my eye on the ball. They include Adelaida Lopez, Karen Brumer, Henry Brumer, Irene Brumer, Joseph Slusky, Katie Hawkinson, Dicky and Marshall Swedlow, Tim Pylko, and Leanne Shigemoto.

Introduction

Golf and Technology: Making a Fickle Game More Friendly

I've always appreciated the great Ben Hogan's characterization of golf as "a fickle mistress." Evidently even the man considered the greatest ball striker of all time knew that just when you think you've mastered this sport, it usually dissolves right before your eyes, as if to remind us that human perfection simply isn't possible. Even so, it's a credit to human imagination, perseverance, and the passion for a game that since Hogan's day, golf equipment designers and manufacturers continue to develop technology that results in higher, straighter, and longer shots, truer putts, and a measurably less-frustrating day on the links for golfers of all ages and ability levels.

Even so, it's not all par or better in golfing paradise. For example, although nobody would dispute that contemporary golf clubs are superior to those of years past, the mind-boggling number of equipment choices available to golfers today often make it hard for people to know which products will best fit their individual games. In fact, playing with clubs that don't fit can do a golfer more harm than good, score- and enjoyment-wise. So knowledge of the technology used in today's clubs becomes essential both for golfers to make well-informed buying decisions and for them to get the most out of their games.

Now for the better news. When Arnold Palmer turned professional in the 1950s, if equipment-savvy players found that their irons' lie angles were a little too flat or too upright, they would bang their clubs a few times on the ground until they bent them into shape. Today, sophisticated and often lengthy iron-fitting sessions conducted by golf pros trained as clubfitters, and vastly improved manufacturing techniques help all golfers find a set of irons built to precisely fit their swings.

Driver fitting now utilizes truly cutting-edge technology, as golfers hit shots during fitting sessions under the watchful eye of computerized launch monitors. These suitcase-sized computers use high-speed cameras or lasers to record a plethora of impact data, such as clubhead speed, the angle at which the ball is launched from the clubface, and the rate and direction of spin imparted on the ball. From this information, trained clubfitters can prescribe a well-fitted driver, with the precise formula of correct length, sufficient loft, adequate shaft flex, and a grip size "molded" to the hands' size to maximize each golfer's potential. Thanks to technology, then, golfers today should expect nothing less than a set of golf clubs that fits them as well as a tailor-made suit or an expensive pair of shoes.

It is in itself a bit mind boggling to consider just how complex and sophisticated golf equipment has become today. One of the goals of this book then, is simply to examine, explore, and explain (and celebrate!) the most advanced expressions and applications of technology in all areas of golf, with the specific goal of helping golfers utilize technology to both play golf better and enjoy it more.

Back to Golf's Future

Let's take a look back to where this boom in technology began. Golf equipment in the later part of the twentieth century saw tremendous advancements in technology, thanks to the imagination and genius of three golf industry executives who, sadly, are no longer with us.

Karsten Solheim

During the 1960s, a folksy engineer named Karsten Solheim revolutionized golf equipment with his investment-casted PING putters and irons. Solheim took up the game in his 40s and quickly realized that the quickest route to lower scores would be to improve his putting. So he thought of a tennis racket, where wooden frames encircled and structurally supported the strings, and realized that the same principles could be applied to golf clubs. Solheim made the first PING putters by placing extra weight on the heel and toe of his clubs for stability and forgiveness on miss-hits, and in so doing he started the passionate search for ever-more-advanced game-improvement, perimeter-weighting strategies for golf clubs that continues to this day.

Gary Adams

Next, PGA golf professional Gary Adams founded TaylorMade Golf after convincing himself that he could drive the relatively new two-piece, solid-core distance balls farther with a metal-headed driver than with a wooden one. TaylorMade's cast perimeter-weighted metal woods, which made their debut in the late 1970s, not only drove the two-piece golf balls further, but because they could be easily and consistently mass produced, they simultaneously supplied and fueled the burgeoning boom in golf. It should be noted that today, virtually every quality golf ball on the market is a solid-core ball and that it was the introduction and proliferation of metal woods that directly led to the technological advances and improvements in today's golf balls as well.

Ely Callaway

Finally, in 1991, Ely Callaway, a businessman, entrepreneur, and distant cousin to golf legend Bobby Jones, took golf and technology to a new level with the introduction of Callaway Golf's Big Bertha extra-large stainless steel driver. Callaway's chief club designer, Dick Helmstetter, developed both a lightweight metal and the manufacturing casting technology to thin the metal walls of that club, which produced both an expanded "sweet spot" for unparalleled forgiveness and a thinner, "hotter" face for extra distance.

Certainly, without the triumvirate of Solheim, Adams, and Callaway, golf would not have enjoyed its Renaissance of technology or grown into what is arguably the most popular participant sport in the world today.

In fact, one can argue that today's new golf club technology does little more than further advance the breakthroughs and contributions made by these three golf equipment Hall of Famers.

A Quick Look into Hi-Tech's New Bag

But just what are these new technologies, and how can golfers best take advantage of them? In a manner of speaking, golf clubs are like the sky or the sea: There's so much more to them than meets the eye. For now we'll just stargaze a bit and touch on the topics that glow the brightest before diving in full fathom five in the chapters to come.

The place to start has to be the hottest topic of them all: today's new large thin-faced titanium drivers, revved-up as they are right to the United States Golf Association's COR (coefficient of restitution, sometimes referred to as the "trampoline effect") limit of .830 for distance. Simply put, this means that if a ball were struck with a driver at what has been the USGA's test speed of 109 mph, the energy transferred from the club to the ball would not exceed 83%. On impact with the ball, these drivers' clubfaces deflect just a little, and this flexing allows the ball itself to retain a little extra energy, which results in longer shots.

Speaking of golf balls, the technological advances they have made in just the past two years has been truly astonishing. Today's crop flies higher, travels farther, spins better, and the balls are more durable than ever before. In addition, ball companies have come up with a host of different types of designs, each using technology to engineer different performance characteristics. In the pages that follow, some of this magic is exposed, so that every golfer will be able to find the right ball for his or her ability level and style of play.

A wonderful new "animal" has recently arrived on the iron scene. Alternately called a combo, hybrid, or blended set, it integrates various degrees of cavity-back and muscle-back irons into a set of clubs. Nike Golf's new Forged Pro Combo Irons, for example, feature full-cavity-backed long irons, which helps get the ball well up in the air, half-cavity mid-irons, which blend forgiveness and control, and classic blade short irons for maximum shot making and dead-aim accuracy. Low-tech isn't no-tech, however, and classic forged muscle-back blades continue to stage a comeback. At the other end of the iron spectrum, designers are already busy making even more forgiving and easier-to-hit, game-improvement irons, and not just for beginners and high-handicappers. One of golf's enduring and, perhaps, most charming ironies is that, often, the game's best players find they too get a boost from products designed for far less accomplished players.

For years shafts have been called "the engine of the club," because they produce the speed, flex, and kick needed to propel the golf ball well into the air and accurately toward the golfer's target. The 1990s saw a number of technological advancements and innovations in graphite shafts that manufacturers have continued to improve upon to this day. The biggest improvement in shafts involves both the materials from which they are made as well as their manufacturing process. Companies now make their

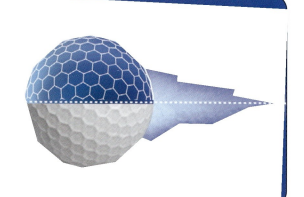

shafts out of space-age-quality graphite, usually combined with boron or other exotic metals or materials.

Although the shafts of the past look from the outside a lot like the shaft of today, new putters have evolved with a visual distinction all their own. Witness Titleist's Futura putter, fresh from the studio of master putter designer Scotty Cameron, which looks like a pancake flipper used by an extraterrestrial. Actually, this putter—which can be classified as a large, rear-weighted mallet—has its (very recent) precursor in the amazingly popular White Hot 2-Ball putter line from Odyssey Golf. Those clubs feature clubheads with two flat concentric white discs the

exact diameter of golf balls positioned on top of the putters' clubheads. The theory (which, refreshingly proves itself true in practice) behind these and other large-headed mallets, such as Nike Golf's Blue Chip Oz and Ben Hogan's Big Ben, is that the clubheads' depth makes it easier to align the clubfaces to the target.

This book doesn't overlook technological advances in wedges, grips, utility clubs, and fairway woods, either.

Titleist's Futura putter

Teching Your Way Around the Golf Course

Golf isn't played with equipment alone: You need a golf course on which to use all of this great stuff. Not surprisingly, when we get to the golf course itself, we find a whole new host of technological innovations that have made the work of golf course architects and superintendents infinitely easier. For example, architects use computer-assisted design (CAD) technology to whip up three-dimensional topographical landscapes to give them a clearer picture as to how they want to shape their land into a course. Computers then make exact calculations as to how much dirt needs to be moved and to where on the property it needs to go. Because it can cost as much as $1.75 to move a yard of dirt (and some courses need to move close to 2 million yards during construction…we're not talking about playing around in a sandbox!), such efficiency in planning represents considerable savings to the golf course developer(s).

Technology has transformed the profession of caring for golf courses as well. Whereas in the past, golf course superintendents had to ride a golf cart to different parts of their courses to adjust their courses' irrigation systems, today wireless handheld remote-control units bring their entire facility's watering needs and operation right to the superintendent's fingertips.

When it's finally time to play a round of golf, pinpoint-precise global-positioning satellite (GPS) systems on golf carts navigate golfers from tee to green with the accuracy and precision of a NASA Mars shot. Golfers who prefer to walk the course can purchase a compact GPS device that they can hold in their hand and carry in their pocket. Speaking about playing, golfers can now reserve their tee times on courses anywhere in the world through a host of online booking agencies.

(High-Tech) Practice Makes
(for Virtual) Perfect(ion)

Let's listen to Ben Hogan again, who, although a man known for his sphinx-like inscrutability, had a lot of wise things to say about the game of golf. When asked to share the "secret of golf," Hogan tersely replied: "The secret is in the dirt; go and dig it out of the dirt." Of course he meant practice. Since the 1990s, golf has seen the proliferation of practice training aids and devices in every odd shape, form, and modus operandi imaginable. Although some of these things bind and harness themselves to golfers in ways that in another context would challenge most people's sense of propriety, they definitely have played a role in improving many golfers' technique. One of the more intriguing applications of technology in this genre is a video system that uses sensors attached to golfers to create a digitally rendered out-

line of a golf student's swing. The instructor then superimposes this animated image over video of his or her student swinging a club. After donning a pair of 3D virtual-reality glasses, the student sees in 3D a virtual display of his or her golf swing inside a digitally animated swing. Other remarkable applications and examples of virtual golf are here or on their way, and we'll go looking for them in this book.

Whereas golfers have always dreamed of hitting their shots on the "line," they are now jumping online to take long-distance lessons from teaching pros with interactive web sites. Sports psychologists who specialize in golf have released CDs that repeat New Age mantras through the headsets with the intention of helping golfers connect to that ever-ephemeral and elusive plane of optimum "here-and-now" performance otherwise known as "the zone."

You may find yourself asking, "Is this just good fun? Or has the world gone totally bonkers over a silly game?" Probably a little of both.

Swing Solutions' GVA 600—an example of digital motion analysis

From the Preppy to Techy Look

Finally, let us not forget the opportunities for sartorial splendor that golf presents to men, women, and children alike. Indeed, as golf apparel has become more hip and chic, in many cases technology has turned clothing into bona-fide pieces of golf equipment as well. There are golf shirts made of microfibers that whisk away perspiration on humid summer days. Designers have taken the awkward bulkiness out of the rain gear of years past and replaced it with sleeker-fitting designs engineered to allow golfers to swing freely in inclement weather. Golf shoes flex and offer support for the feet in ways that work in harmony with proper footwork during the swing. Golfers can now switch the interchangeable lenses of their sunglasses for better depth perception depending on the degree of glare. Golf bags have gotten lighter yet at the same time have more room in them for everything but a reading lamp and a best-selling novel.

Sunglasses from Oakley

Golf's Essential Paradox

Indeed, at the end of the day, what is so engaging about golf may be its essential, mysterious, and enduring paradox. On the one hand, you have the most elemental and primitive game imaginable, whose modus operandi is just to hit a ball with a set of sticks over a field of grass into a series of holes. On the other hand, golf administers a sophisticated test of a person's strength, flexibility, endurance, capacity for concentration, and threshold for frustration, which when successfully negotiated yields a sense of satisfaction and pleasure unlike any other on Earth. I like to think of technology as a bridge that integrates these "two hands" into one complete human being, and in so doing frees people's hearts and minds to truly play golf. If I am right, this may be the game's greatest gift of all.

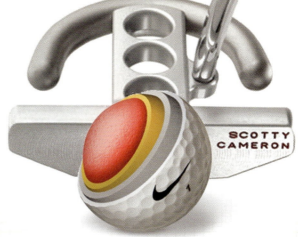

The
Golf Gear

"A PGA Tour pro's equipment is as important to him as a dentist's tools are to a dentist."

—David Ogrin, PGA Tour tournament winner

Technology is a tool, a branch of science applied to make the tasks and pleasures of everyday living easier, and in some cases, more fun. Therefore golf clubs are simply tools, whose purpose is to help golfers get the ball into the hole in as few strokes as possible and with as much pleasure as the devilishly frustrating game of golf allows. In this section, we'll look at the different types of golf clubs (drivers, fairway woods, irons, utility clubs, wedges, and putters), as well as golf balls, golf shoes, gloves, and even bags, because they too aid in helping golfers play better and enjoy the game more.

While all golfers recognize immediately when their equipment functions up to par, not everyone knows the amount of time, effort, imagination, trial and error, and, most importantly, passion that equipment makers invest when making their new products work better than their precursors.

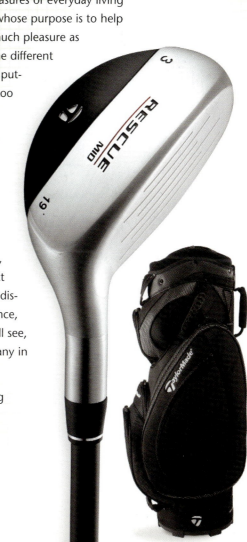

In fact, technology as it is applied to golf club and ball design is a wonderful metaphor for the spirit of compromise and moderation that sages since Socrates have identified as a key to a more satisfying life. For example, today's large titanium driver heads make it easier to make consistent contact with the ball, but challenge designers to position the club's internal weight distribution properly. A golf ball that spins less off the tee promises extra distance, but less control when using an iron and when around the green. As we shall see, the golf ball designers' solutions represent a design breakthrough equal to any in the evolution of golf equipment.

Let's start at the beginning of our round by pulling the driver out of our bag and banging one long and straight down the fairway.

The Driver: COR Hits It Far

The driver is the most glamorous club in the bag, not only because it hits the ball farther than all the others, but also because watching the extended parabola of a golf ball in flight offers a poetic thrill like few others in sports. Back down on Earth, driving the ball well sets up a golfer's subsequent shot or shots to the green and is essential for good scores. Even Tour players, who have lucrative contracts with golf club equipment makers to use their products, insert language in their contracts that allows them to play the driver of their choice, even if it's from a competing company. In other words, the importance of driving the ball both long and straight is inviolable and incontestable.

Club manufacturers have revved up virtually all of today's crop of thin-faced titanium drivers to the USGA's coefficient of restitution (COR) or "spring-like effect" limit of .830 for ball speed. Simply put, COR is the transfer of energy between the golf club and the golf ball during impact, which can be measured in the speed of the ball as it leaves the club. During the USGA's present COR test, a golf ball is fired into a stationary driver at 109 mph. The ball's speed when leaving the clubface cannot exceed 160 mph; if it does, the driver has an illegal amount of COR. In the future, the USGA will be using a different test.

Drivers from
Callaway and Mizuno

Dick Rugge, Senior Technical Director, USGA, on the new USGA test for drivers' COR

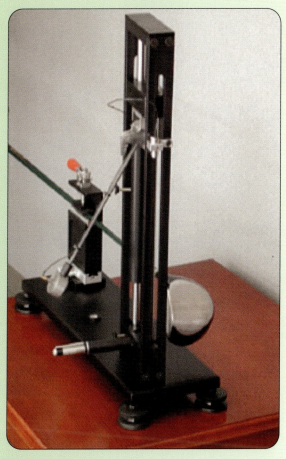

The new USGA test

The COR test device we have been using is pretty cumbersome, because the equipment we use is large and the test is time-consuming. So we came up with a simplified and quicker way of testing for a club's spring-like effect using a pendulum device. On the original COR test we had to disassemble the clubhead from the shaft. On the new test, we don't have to do that; we can take a finished, meaning a whole complete club, and test it.

The new test attaches a metal mass at the end of a pretty short pendulum, about 10 or 12 inches long, which swings into a stationary clubface. There's no spring in it, this metal mass just swings by gravity. We have an accelerometer, which is an electronic device used to measure vibrations, attached to the mass. During the collision between this mass and the clubhead, vibrations are created, which enable us to measure the time that this mass stays in contact with the clubhead. That time corresponds very accurately with the COR reading that we get within the COR test.

Actually, we're not going to be measuring COR any longer, we're going to be measuring time. The initial limit that we have proposed of time this mass can stay in contact with the clubface is 250 microseconds. If it stays in contact longer than this, the club will not conform to the new rule. This new test is a more direct measurement of the springiness of a club, rather than an indirect measurement of the collision between the ball and the club in the COR test. Just like with the COR test, however, the springier the clubhead, the faster the ball comes off that face and the farther the shot will go.

Today's high COR titanium drivers usually have very thin clubfaces forged from high-grade and very strong beta titanium. This material allows the clubface to flex or deflect, then spring back on impact with the ball without breaking. When a golf ball is struck by these clubfaces, the faces compress and then rebound, and push the ball away with added speed. That's what it means for a driver to have a high COR. What's more, because of the way the clubface deflects during impact, the ball itself deforms or compresses less than it would against the more rigid clubfaces of years past. As a result, the ball loses less energy in this collision, and combined with the titanium clubface's action of springing forward, the ball flies down the fairway at prodigious lengths. Although this sounds like pretty heady science, suffice it to say COR hits it far.

But it's the medium that really delivers the message as far as today's titanium drivers are concerned. Titanium—the medium—is an extremely strong and lightweight material, which means club designers can build clubs both very large and with a great deal of leeway to position the weight on the insides of the clubheads exactly where they want it. In fact, weight manipulation and distribution is what contemporary golf club design (all clubs, not only drivers) is all about.

Actually, the large size of today's oversized titanium drivers (most are between 350 and 460cc [cubic centimeters]) accomplishes two other important performance goals: It makes the drivers springier, because the more the titanium is stretched out, the more rebound effect it is capable of producing; and it makes them inherently more stable and forgiving on off-center hits (just as if a house were struck with a thrown baseball, it would wobble far less than a frying pan struck with the same pitch).

It must be pointed out, however, that golfers with high swing speeds, say 100 mph and more, benefit most from today's high COR drivers, because it takes at least this much speed to compress the ball against clubface to make the spring-like effect work to any significant degree. But, as Benoit Vincent says in his sidebar later in this section, the next frontier in terms of high COR drivers will be their consistency, not additional distance.

Today, leading driver makers such as Callaway Golf, TaylorMade, PING, NIKE, COBRA, Mizuno, Adams, and others design their driver heads with the help of sophisticated computer-assisted design (CAD) programs. With the push of a key and the sliding of a mouse, club designers position weight inside the clubheads to increase the clubs' stability and forgiveness at impact. Not only that, but they can produce prototypes at a rapid clip, which means they can test the effectiveness of their weight placements, and then just as easily go back to the CAD drawing board and revise as needed. The use of graphite "swatches" positioned on the driver's crown is, perhaps, the newest weight-saving technology in driver design. Callaway, Mizuno, and Yonex were the first to introduce such drivers. Not only does the lightweight graphite enable designers to reposition the saved weight in other parts of the club (lower and deeper in the clubhead is their usual choice), but it also flexes the entire clubhead slightly on impact for a slight catapult effect that propels the ball farther.

However, the human eye and hand remain important parts of the design process. Tom Stites, Director of Product Creation for Nike Golf, believes that a well-designed golf club is never a matter of 100% technology and that there is always art and craftsmanship involved. He points out that there is a crafts-man with a file in his hand next to his company's CNC/CAD design station (where designers develop and refine club prototypes) who blends and sculpts into working molds the sub-tleties of the club's shape in a way that only human sensibility and sensi-tivity can.

Finally, an innovative company called Burrows Golf has focused its attention on a different aspect of the energy generated when club meets ball. Their MAC driver has a hollow inverted dome cast out of the club's soleplate called the Powersphere. The company contends that when the club strikes the ball, energy shockwaves surge toward the back of the clubhead. There they hit this Powersphere, which reflects them back across the entire face of the club, which amplifies the power transmitted to the ball. Although golfers will have to empirically test and decide for them-selves if they hit longer and straighter shots with this driver, even golf writer veterans who claim they have heard it all find the theory, at least, behind this tech-nological story novel and refreshing.

The MAC driver's Powersphere from Burrows Golf

TechTip

Because the large titanium drivers with high COR position their centers of gravity (CG) low on the clubface, it is important that golfers contact the ball above this position. To do so, tee the ball a little higher than you may be used to, so that the top or crown of the driver is even with the bottom of the ball.

Benoit Vincent, Chief Technical Officer, club and ball research and development for TaylorMade-adidas Golf on drivers

The most important thing about drivers is their perimeter weighting, which is a "weight management" problem that asks the question, "Where do I put my weight in the club to make it perform at its best?" You keep trying to make the shell or body of the driver lighter and lighter and lighter, which lets you build clubs bigger, which in turn can allow for more weight-distribution possibilities. Today we're making drivers larger by using titanium, and the trend as we go into the future is to find even lighter materials, perhaps composites. The quest, however, for new material is always driven by the need to manipulate weight differently in the head.

Can we make the titanium even thinner so we can make the drivers bigger? The challenge with a 450 or 500cc head is that you have to keep the same amount of material in the clubface so that it won't collapse during impact, and that means you will have a heavier face in proportion with the rest of the club. This moves the center of gravity on such a huge driver forward, and a golfer will tend to hit shots with that club that fly low and slice to the right, which isn't good.

The other big battle is to preserve the benefits of COR on off-center hits, because although these big clubs allow you to easily make contact with the ball away from the center of the face, they also lose ball velocity, thus distance, when you do. We solved this problem to some degree on our R500 series drivers by using what we call "inverted cone technology."

What we've done is taken an additional piece of titanium and milled it to a specific size and geometry (the shape of a cone), which we then weld with the clubface to the body of the driver. This cone has a high COR, so it expands the COR zone to a wider portion of the clubface. We have to acknowledge that we have maxxed-out our distance capacity with the driver because of the COR limit of .830, so when we talk about drivers in the future, we're not just talking about increasing distance. What we will be working on will be the issue of the drivers' inconsistency. The inverted cone technology is one way to engage the COR zone more consistently, but there will be others.

The inverted cone technology

Billy Mayfair, PGA Tour player on the technology of the driver

Billy Mayfair

The technology has gotten unbelievably good in today's drivers. The bigger their heads become, it seems the easier they are to hit. The biggest change due to driver technology, however, is that now you can swing harder at the ball and not worry about hitting your drive too far off line. In fact the ball seems to correct itself, or straighten itself out in flight. If you go back to the old persimmon heads, the wooden heads, and you misshit a shot, you really misshit it badly. With these new drivers it seems that when you don't hit the ball the way you want to, it still flies pretty well.

Now I've changed my swing a little to adjust and to take advantage of the new equipment. I had more of a long fluid golf swing growing up, and now it's a little shorter. I've worked to build more clubhead speed into it to hit the ball longer so I can compete on the longer courses we play on the PGA Tour. In other words, the technology today is designed to hit the ball higher and farther, so you've got to change your game to go along with that technology.

Arnold Palmer, World Golf Hall of Fame member, on today's titanium drivers

Arnold Palmer

The oversized titanium drivers of today have prolonged the life of a lot of golf pros, and have given them an opportunity to play a lot longer than they might otherwise, myself included. I can still get some reasonable distance out of hitting a driver, and technology is the reason. I think the technological advancements in golf equipment have created a new interest in the amateur player, too.

Fairway Woods: About Face

In recent years, one of the best golf tips offered by teaching pros had nothing to do with the grip, or the stance, or even the swing. Instructors just told golfers to replace their long irons with easier-to-hit fairway woods. The larger heads of modern fairway woods make them easier to hit off the tee, while better sole designs means they glide through the rough more efficiently than ever. Tiger Woods popularized the fairway wood chip shot, a technique unimaginable with the smaller-headed fairway woods (when they really *were* woods) of years past. As more and younger PGA Tour players used fairway woods to attack the tightly tucked pins on par 5s, par 4s, and par 3s, amateurs finally followed suit.

In 1993 Adams Golf's Tight Lies line of fairway woods changed the thinking about these clubs' designs. In fact, prior to Tight Lies, there *wasn't* a lot of thinking about these clubs: People thought of them as "little drivers" that they hit from the fairway. The Tight Lies changed that with their low-profile/shallow-faced clubheads, whose low centers of gravity made it easy to hit the ball high in the air. Orlimar soon followed with their similarly configured TriMetal line, and the shallow-faced fairway metal has been a golf equipment staple ever since. What's more, these clubs reminded clubmakers that less-skilled golfers, older players, and players with slower swing speeds needed help getting the ball up in the air. Starting in the early 1990s, they began marketing the kinds of high-lofted 7, 9, even 11 woods that have improved games and helped countless golfers enjoy the game more.

Adams Golf's Tight Lies started a trend in shallow-faced fairway metals that were easy to hit and assisted in getting the ball well into the air.

Barney Adams, founder of Adams Golf, on the invention of the Tight Lies fairway woods

Barney Adams

The Tight Lies came about in 1995 as a result of my experience as a custom fitter, because the people I was fitting kept asking, "Will these custom-fitted clubs help me hit my long irons better?" So I could tell that those players were really tired of struggling with the hard-to-hit long irons. Now this was in the early to mid-1990s, the first era of the oversized drivers, and fairway metal woods were getting larger then, too. I saw my customers struggle to get the ball up in the air with these big fairway woods, and I thought to myself, "This is silly, I need to design a fairway metal that will help them hit the ball higher." And that was the motivation behind the first Tight Lies.

Now if you look at the design of a traditional driver or metal wood, you'll see that it is diamond-shaped. The clubhead is broad across the top and narrows at its sole plate. I looked at it logically and said, "If I want to move the weight down on a fairway metal, I have to take the diamond and turn it upside down." Now, the wider part with most of the weight in it will be low toward the bottom of the club where you hit the ball.

I also shallowed the face of the Tight Lies considerably, because I knew that a good design has to instill confidence in the golfer. I wanted people standing over their shots to feel that they could get the ball up in the air easily with this club, without getting tight or trying to muscle the ball up in the air. In other words I wanted people to release the clubhead, and by making the club shallower, they didn't feel like they had to help the ball up. They could make a nice smooth swinging motion through the ball.

Although shallow-faced fairway woods did wonders for high-handicappers, better players with faster swing speeds often hit the ball too high with these clubs. In addition, although Tour players and better amateurs could hit much farther with their large, high-COR drivers, their missed shots were also flying farther into trouble. Consequently, they looked for strong-lofted fairway woods to play from the tee for more control. The shallow-faced varieties proved ill equipped for this task, however, because their clubheads swung too far under a ball placed on a tee, and they tended to pop the shot straight up into the air.

The large-headed titanium fairway woods presented different playability problems. The cumbersome mass and height of these clubheads wouldn't glide easily through the roughs, and the clubs' centers of gravity sat too high on the clubface, which resulted in shots hit too low.

To make these larger and deeper-faced fairway woods work well, designers had to do something to lower their centers of gravity. The solution came when improved casting technology allowed clubmakers to design lightweight, mid-sized fairway woods with thinner and springier faces and walls. The thin faces increased the club's COR for extra distance, and the thin walls gave designers more internal perimeter weighting options for greater stability.

To lower the centers of gravity in these fairway metals, companies such as Nike, Adams, Tourstage, Yonex, PING, Wilson, and others began positioning heavy tungsten plugs (or pads weighted with different materials) low and rearward on the sole plates of these clubs, which instantly produced higher shots. Golfers now had the best of all worlds in fairway woods with larger faces, with which they could contact the ball easily, and with low and deep centers of gravity for higher shots.

To meet the needs of golfers of all ability and strength levels, companies began offering lines of fairway woods that included both shallow and taller faces, that corresponded to lower and higher face weight placements, respectively, for players with slower and faster swing speed.

Nike Golf's T-40 features a stainless steel clubhead, with a 40-gram tungsten plug positioned toward the bottom of the clubhead to lower the center of gravity and increase the ball's trajectory.

TechTip

"Hit down on the brown," golf legend Gary Player wrote in an article on fairway wood play. By that he meant the golfer should strike fairway wood shots with a slightly descending blow that will leave a shallow brown divot of earth after the ball is on its way. Clubmakers have facilitated the task of hitting down with fairway woods by improving the designs of the camber and shape of their clubs' sole plates so that they skid or glide smoothly, rather than dig and twist off line, as they contact the ball.

Sonartec's "driving cavity" positions the club's center of gravity even with the middle of the ball, which prevents shots struck with high clubhead speeds from ballooning into the air.

Recently, Sonartec, one of the newer equipment companies, saw an opportunity to concentrate on fairway metals for the highly skilled player, because these golfers tended to hit even the deeper-faced tungsten weighted fairway woods too high. Their solution came through a clever clubhead design called the Driving Cavity, which removed a chunk of metal from directly behind the clubface. The center of gravity on this moderately deep clubhead wasn't too high or too low, but rather aligned itself directly with the center of the ball. Stronger players could hit these fairway wood shots as hard as they wanted, without worrying about the ball ballooning too high into the air, and could also play them from the tee. Sonartec offers clubs for all players, and rounds out its line with their shallow-faced and semi-deep faced models.

Irons: The Weighting Game

Irons are the unsung heroes of a golfer's bag. Perhaps because there are so many of them (up to 13, if counting the wedges), their yeoman-like task of delivering the ball close to the pin on hole after hole fades behind the grandeur of a long drive and the suspense of a long breaking putt rattling in the bottom of the cup. When people speak of the game's great ball-strikers, such as Ben Hogan, Lee Trevino, and Annika Sorenstam, however, they are, even if unknowingly, praising these players' prowess with the irons. Because the irons vary in length and lie angle through the set, they require the kind of subtle swing adjustments only the skilled player can make. So to become better all-around players, golfers should spend plenty of their practice time working on their iron game. Indeed, technological improvements in irons have allowed golfers to improve this segment of their games faster and with more success than ever.

Each category of irons is now more consistent and easier to play. Blade irons, for example, offer new subtle weight-distribution strategies that add just a bit of forgiveness to these small-headed, therefore classically "hardest-to-hit," clubs. Larger perimeter-weighted cast irons, whose size in years past made it difficult to maneuver the ball, have trimmed down and now display greater shot-making capacities than before.

Adams Golf's Idea Irons

As blades and cavity-backed irons move closer together, a new way of imagining and creating a set of irons has emerged. Alternately called hybrid, combo, mixed, and even morphed sets, the grouping consists of full-cavity-backed long irons, half- or partial-cavity-backed middle irons, and classic muscle-back blade-style short irons and wedges. Companies such as Nike, MacGregor, and TaylorMade all make forged hybrid sets (Adams Golf's Idea Irons represents a cast version), which integrate these classes of irons into unified and visually integrated sets. But hybrid irons represent a new choice, not a mandatory prescription, for golfers looking to improve their iron play: Many golfers still opt for full-blade or cavity-backed sets of irons, and these have improved considerably in and of themselves in recent years.

Traditionally, forged muscle-back blades appeal to the pro or low-handicap amateur player. Their smaller clubheads, with weight positioned relatively high directly behind the club's center, or sweet spot, as well as thin soles and sharper leading edges all allow skilled golfers to carve their shots with fades or draws, and hit the ball high or low at will. Blades offer superb feedback to the golfer as well, which the skilled player wants, so he or she can feel and analyze miss-hits and then make the appropriate corrections in the swing.

Technology has made these difficult-to-hit clubs a little easier to play, even for the pros. For example, Ben Hogan Golf's latest Apex blades feature a weighted oval-shaped portion of metal that extends up or down on the back of each clubhead depending on the iron. The taller pad raises the center of gravity on the short irons, which keeps shots from ballooning too high, and it recedes on the long irons, which lowers the clubs' centers of gravity to hit the ball high. Mizuno, known for their premium and pure blades, offer several models with different performance characteristics. The MP 33, for example, features a lower center of gravity for high shots, whereas the MP 37 moves the "CG" up a bit for more mid-trajectory shots.

Mizuno's MP 37

Blades enjoy an unquestioned status as the best feeling of the irons, although TaylorMade has developed technology to make them feel even more pleasing to the hands. The company's RAC muscle-back irons position small pockets of metal on each clubhead, which channel the shots' vibrations in a preprogrammed manner for a consistent feel between irons throughout the set.

Aesthetics remains an underestimated aspect of irons (except by the best players), but aesthetics are important because a set that pleases the eye can also instill confidence in the mind. CAD technology has allowed clubmakers to reproduce the sophisticated blend of shapes and angles that comprise a set of irons much more quickly than ever before. The exact coordinates of a CAD-designed iron head allows companies to manufacture set after set of irons from virtually identically shaped molds. In years past, each new mold would present small but not insignificant variations in an iron's head, which resulted in potential performance inconsistencies from set to set.

Benoit Vincent, Chief Technical Officer, club and ball research and development for TaylorMade-adidas Golf, on CAD-designed irons

Benoit Vincent

Although not all companies like TaylorMade use CAD to design their irons, those that do can manufacture consistent products set after set. Let me take you through a basic step-by-step tour of how club designers use CAD to produce a set of irons.

The first thing the designer does is to enter or "place" a set of coordinates of lines and surface into the computer, which reproduces or "draws" the shape of the iron head we want on the computer screen.

When we are happy with the shape visually and aesthetically on the computer, we move to the club's weighting. The computer calculates the overall weight of the iron and where the weight should be placed to achieve the center of gravity and inertia, which gives us the stabilizing qualities we want.

When that is done, we push a button, and the data is sent to another computer, connected to a CNC machine (computer numeric control) that follows the numeric path provided by CAD and cuts what we call a mock-up iron out of lightweight resin. Now we have a real 10- or 20-gram 3D mock-up of our iron, and this lets us see all the defects of our iron's shape, so we can revise and revise the iron as many times as necessary, making sure each change doesn't affect the club's weight. Of course, this process is repeated with all the irons in the set, because each club has its unique specifications and weight-distribution requirements.

When we have the iron designed precisely as we want it on CAD, we again turn to the CNC machine, which this time carves a master out of a copper block. From this master we make two sets of tools, one for cast irons and one for forged irons, from which we mass produce sets of clubs.

Cast stainless steel perimeter-weighted irons have improved considerably in recent years as well. Generally designed for the middle- to higher-handicap player (although also used by some Tour pros), these clubs offer lower centers of gravity, which makes them easier to hit high, and cavity-back designs for added forgiveness on off-centered hits. CAD technology and improved casting methods have allowed clubmakers to efficiently remove metal from one part of the club and reposition it elsewhere to improve performance. Callaway's Steelhead X-16 irons are an excellent example.

Callaway's Steelhead X-16 irons

The set's first weight-watching measure comes via the company's legendary S2H2 reduced hosel design, which saves weight by shortening the sizes of the irons' hosels, or the top part of an iron into which the shaft is inserted. Next, each iron has a 360-degree undercut channel, or a thin slot of metal removed from directly behind the clubface, which moves weight back in the clubhead for higher shots. Finally, designers have removed a "notch" or chunk of metal from the back bottom of the clubheads and repositioned it around the club's perimeter for added stability and forgiveness on off-centered hits. Other models of cast cavity-backs from Cobra, TaylorMade, PING, Mizuno, and other companies use different design strategies, but all strive to achieve higher shots, more stable clubheads, and an attractive look and good feel.

Perimeter-weighted versions of the avowed hard-to-hit long irons traditionally use maximum perimeter weighting for added stability. They also have the lowest and deepest centers of gravity for higher shots. Irons such as Callaway's Hawkeye VFT irons achieve these ends via lightweight forged titanium clubfaces, which are attached by compression to lighter, thin stainless-steel clubheads. Remember, this removal of weight from the clubface enables designers to reposition the weight precisely where they want on the clubhead. Titanium-faced irons may feel more harsh at impact than forged irons

and cast cavity-back irons, but many golfers, especially higher-handicappers, gladly will opt to pay this price for better shots.

Club designers have started to develop entire iron sets such as Nike's SlingShot irons with large, stable, and easy-to-hit heads that closely resemble the utility, hybrid wood irons that have become extremely popular. (Although mostly meant for beginners and high-handicap players, these will no doubt also find their way into better players' bags.)

Finally, steel shafts for irons keep getting lighter and lighter. Whereas even 2 or 3 years ago the standard "off-the-rack" iron weighed close to 120 grams, today several are under 100 grams. Jack Nicklaus has always chosen to play with relatively lightweight irons, and the golf industry, it seems, has finally followed this champion's lead.

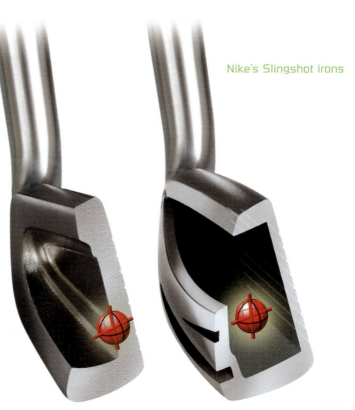

Nike's Slingshot irons

Justin Leonard, 1997 British Open champion, on irons

As a forged-blade player, I really don't embrace all the technology available in irons today, as golfers who play cast, perimeter-weighted irons might. But that's not to say I don't appreciate the improvements in blades. Mainly what I'm looking for in my irons is consistency and feel from shot to shot, as well as the capacity to work the ball, which means trajectory control and maneuverability. In other words, I want to be able to easily work the ball from right to left or from left to right, and forged blades let me do that best. Last, but not least, my irons, like my wedges, driver, fairway woods, and putter, must appeal to my eye, because this gives me a feeling of confidence.

Justin Leonard

Mike Ferris, VP of marketing for Ben Hogan Golf, on the future of irons

Although it's true that hybrid or mixed sets of irons are emerging on the equipment scene, some golfers have been mixing and matching their sets for years. At Ben Hogan Golf, for example, for quite some time players have been able to create their split sets by selecting the combination of clubs they want from our Apex blade, perimeter-weighted Edge Pro, and oversized Edge CFT model lines.

In the future, I think you'll see not only more of what I like to call morphed or split sets of irons, but also a range of these sets. By this I mean that there will be blade-like sets designed for greater workability and precision for the pros and the low-handicap players, and sets that take what are now called game-improvement designs even further for less-skilled players. In these, each iron in the bag will have a unique design as well as technological features that transition in a gradual way from one club to another.

Utility Clubs: Rough Riders

Legend has it that two of sport's all-time greats, Ted Williams and Sam Snead, were debating the challenges of their respective games. Williams said that hitting a baseball pitched at 90-plus miles an hour was the single most difficult thing to do in sports. Snead quipped, "That may be so, but in golf you have to play your foul balls!" Ben Hogan seconded Snead's sentiment when he said, "In golf, it's not how good your good shots are that matters, it's how good your bad shots are."

Today's utility clubs, or hybrid wood/irons, take Snead's and Hogan's advice to heart. With precursors such as the Baffler from Cobra Golf and the Ginty from Stan Thompson Golf, the new hybrid clubs get golfers out of trouble and back into play. The contemporary lineup of hybrids includes The Rescue Mid Club from TaylorMade, Sonartec's T.R.C, and the Knife from LJC Golf. All power through high rough, divots, and other tough lies to get golfers on the green even from considerable distances away.

TaylorMade's
Rescue Mid Club

One can also trace the idea of the contemporary hybrid back to the shallow-faced fairway woods, such as Adams Golf's Tight Lies and Orlimar's TriMetal. Like hybrids, these clubs position weight low in the clubheads to get the ball up, while their shallow faces let them dig the ball out of difficult lies. Indeed, although the name "hybrid" reflects the way these clubs blend the characteristics of fairway metals and irons, hybrids differ from fairway woods in significant ways.

First, their similarities. Both hybrids and fairway woods are cast from stainless steel. They both have broad, hollow heads and soles that are often equipped with rails or runners designed to interact efficiently with the turf. Both hybrids' and woods' centers of gravity are positioned low and back on the clubs, adding dynamic loft to shots and helping the ball into the air.

The hybrids start to lean more toward the irons, however, because they are designed for distance control and accuracy. Their shafts are shorter than the fairway woods', and their clubfaces are stiffer, with less COR. Although this means less total distance on shots, it also means less distance variation between shots. Now on miss-hits the ball still finishes relatively close to the target. Hybrids' lofts usually start at 16 degrees and progress into the mid-20s, which, again, resemble lofts of the long irons. Their clubfaces are flat and smaller like an iron's, which means that unlike woods, hybrids have little to no bulge and roll and also produce no gear effect.

Simply put, bulge and roll refers to a driver's or wood's slight convex curvature, both up and down and from heel to toe on the clubface. On slightly off-center hits, this rounded clubface surface interacts with a club's gear effect—the counter-rotational action that applies correcting spin on shots hit with

an open or closed clubface—so that shots curve back toward the center of the fairway. Unfortunately, on severely miss-hit shots, gear effect causes wild hooks and slices. Because irons and hybrids have almost no gear effect, shots hit on the toe and heel of the club will fly a little off-line, but will lose less distance compared with a centered hit and they won't curve much further into trouble.

Sole plates are also very important on hybrids. The LJC Knife, for example, presents a radical three-blade sole plate design, with the two outside blades acting as outriggers that stabilize the clubhead as it contacts the ground. The club's middle blade functions like a rudder, which centers an opened or closed clubhead at impact and brings it back to a square alignment.

Sonartec's T.R.C uses two parallel rails that smoothly glide the clubhead over and through the grass, like wheels securely set on railroad tracks. In both the T.R.C and the Knife, rails reduce the clubhead's contact with the ground, which minimizes friction at impact and allows a greater transfer of energy into the ball for more distance. The TaylorMade Rescue Mid club has no rails: Its designers believe that in raising and destabilizing the clubhead off the ground at address, rails make it harder for golfers to line up their shots.

The Sonartec T.R.C combines the driving cavity of the company's fairway woods with a smaller, heavier, and more compact clubhead. It features twin sole plate rails that glide the club through even thick rough for clean contact with the ball. A more upright lie angle enables golfers to attack the ball out of the rough at a steeper swing plane angle, which reduces the amount of grass between the clubface and the ball at impact for more distance and control.

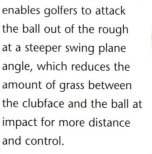

The Sonartec T.R.C

Nick Price, World Golf Hall of Fame, on his involvement in designing Sonartec's T.R.C Utility club

Nick Price   Photo Copyright Paul Lester

Over the last three years I have really enjoyed my relationship with Sonartec. They have always been receptive to suggestions I have made when trying to make their clubs even better. I have always had a passion for golf club design, and have spent most of my life tinkering with golf clubs. Sonartec has given me the opportunity on numerous occasions to help design their new products.

The T.R.C Hybrid is a good example of my involvement with Sonartec. They first showed me the utility club that had been selling in Japan and wanted my input as to what would make it succeed in the United States. I suggested making the clubhead a bit smaller because we wanted to make sure the club could get through long rough and tough lies of every sort, such as divots. As we made the clubhead smaller we also made it a little heavier, to keep the face squarer through high grass and tough lies. The smaller, heavier head required a slightly shorter shaft, so we went to one 3/4 of an inch shorter than the Japanese version. This, in turn, meant we had to design the new club with a slightly more upright lie. As a result, the golfer now stands slightly closer to the ball, which promotes a steeper downswing plane, which helps the ball get up and out of the rough.

Finally, the earlier version had a lot of bulge and roll on its face, so it was more like a fairway wood than an iron. I suggested switching that around so that the new club would have almost no bulge and roll, having a flatter clubface like an iron's design. What we were left with was a club that is extremely easy to use on the fairway, gets the ball in the air like a wood, and is versatile in helping the golfer hit the ball out of poor lies. This is a club that every player should have in their bag, as it makes a wonderful substitute for long irons. I am fortunate enough to still be a pretty good long iron player, but I know it is just a matter of time before the T.R.C ends up in my bag.

Wedges: Great Escapes

Statistics show that during a round of golf, golfers play between 65% and 70% of all their shots from 130 yards or closer to the pin. Thankfully, most pro shops are well stocked with enough wedges to fit every player. Before discussing wedge design and technology, however, it is important to understand the physics of a wedge shot, because with wedges, form follows function closely and cleverly.

The loft of the wedge and its grooves work in harmony to produce the correct spin rate, which enables players to control the trajectory and roll they want on their wedge shots. Not surprisingly, then, companies such as Cleveland Golf, Callaway, Titleist, TaylorMade, PING, Ben Hogan Golf, Wilson, MacGregor, and others offer a number of different lofts (ranging from 48 through 64 degrees) in their wedge lines. But it's in today's advanced groove technology that the discussion of new wedges becomes more interesting.

Much has been written about the geometry of the wedges' grooves, but it really doesn't matter if they are U- or V-shaped. That's because the ball only contacts the first 4 or 5 thousandths of the top part of a groove, where the total depth is 20,000 to 25,000 of an inch deep. The bottom portion of the grooves just channels or throws off the grass, water, dirt, sand, and other debris deposited during impact. What matters, then, is the transition between the flat face of the wedge and its grooves, which need to be relatively sharp and consistent over the clubface to work their best (but not knife-like, because that will shear the surface of the golf ball). Keep in mind, however, that on shots from the grass, a golf ball stays on the face of a wedge for a blazing 0.4 milliseconds, during which the club generates a

Cleveland Golf's wedges

Wedge Fitting

Golfers should make sure that they carry the right combination of wedges, so they don't fall victim to "distance gaps" or shots for which they don't have the right club. If a golfer carries a 49-degree pitching wedge that he or she can hit 115 yards and a 56-degree sand wedge that carries 95 yards, for example, that player may have trouble playing shots in between these distances. The solution would be a 52-degree gap wedge, which the player can hit a maximum of 110 yards, and 100 or 105 yards comfortably. In recent years, the 60-degree lob wedge has become a standard addition to a golfer's bag for pitching the ball high over greenside bunkers or chipping it softly to pins on fast-running greens.

spin rate of 10,000 rpms. By contrast, a golfer swings a driver on a relatively horizontal plane into the ball, which spins very little up the face of the club during impact. Grooves on driver faces, then, are virtually insignificant; in fact, many of the best drivers do not even have them. A wedge shot, on the other hand, finds the ball dynamically sliding up the face of the club where the grooves impart essential shot-controlling spin on the ball. A wedge's grooves have to do their work fast and well.

Grooves in wedges have improved dramatically in the past few years as companies have begun to machine cut them, rather than cast them, into the wedge's face. Casting grooves produces excess metal, which has to be ground and polished by hand, a process that yields inconsistent results. CNC machining the face of each wedge, however, creates perfectly consistent grooves, both on each wedge and from wedge to wedge.

Golfers can pick up a little more spin on their wedges by selecting raw models that have not been coated with traditional protective chrome plating. These clubs begin to rust quickly after use, which creates added friction on the shots. Golfers can buy the feel they prefer by choosing from soft-feeling forged carbon steel wedges, firmer cast stainless steel wedges, or wedges finished with traditional chrome or a soft-feeling nickel coating.

Machined grooves on a TaylorMade wedge

Bounce Tip

Good wedge play means learning how to use the bounce on the bottom of the wedge effectively. Bounce is the angle created by a built-up portion of steel on the club's sole and the club's leading edge. Golfers must match the amount of bounce on their wedges to the sand and grass conditions of the courses on which they play as well as to their styles of swings.

As a rule, harder turf and firmer sand require wedges with less bounce, so that the club's sharper leading edge can dig down and through the shot. Wedges with more bounce tend to work better on courses with softer turf and sand conditions, because these clubs will not dig too deeply into the ground.

Players with steeper or more upright swings generally do better with more bounce on their wedges, because it prevents their clubs from digging into the ground. Those who pick or sweep their wedge shots more cleanly off the turf often select wedges with less bounce, so they won't skid off the ground and contact the middle of the ball in what is called a skulled or bladed shot.

Mother Earth qualifies as the best piece of technology to help golfers know whether they are wedge diggers or sweepers. Just observe the size and depth of your divots after hitting wedge shots. Long and deep divots signal diggers, whereas hallow and thin divots indicate sweepers.

Gary Player, World Golf Hall of Fame member, on wedges and the short game

Gary Player

Wedge play is all about feel, and in this way it stands as a good symbol for the game of golf as a whole. All the knowledge in the world about how to hit golf shots doesn't mean anything until you can imagine them first in your mind and feel, not think, your way through them. When you play those partial shots from the grass or sand, you have to have a wedge that gives you excellent feedback, and, again, that means feel. Design is essential, too, as it works with the wedge's material to give you that ultimate scoring club. I have always preferred my sand wedges to have a nice wide or thick flange along the sole. This increases the bounce, or angle between the leading and trailing edge, of the wedge and lets you really swing down through and most importantly under the sand, so that the ball can pop right up and out of the trap. I've played several different brands of wedges in my career, but several years ago I discovered wedges made from an aluminum bronze alloy that felt like peaches and cream. They were that soft, and even out of tall rough, they put spin on the ball and didn't hit those low-spinning balls we call flyers.

Do you know how much I loved that aluminum bronze material? Well, I bought the company and started manufacturing the Par Saver line of wedges. Now aluminum bronze might not feel as good to everyone as it does to me; forged carbon steel wedges or firmer-feeling cast stainless steel wedges might work better for some. But the point I want to make is that because wedge shots require the ultimate feel to play, play them with wedges that feel the best to you.

Remember that 70% of shots in golf are played from 100 yards in, so practice your short game!

Bob Vokey, master wedge designer for Titleist, on a few tips on selecting the right wedges

Bob Vokey

Work with your pro or go to a fitting center to get fitted for your wedges. Golfers fall into two categories: the diggers and the sweepers. Diggers need more bounce on their wedges, and sweepers less.

Titleist's 56-degree sand wedge

If you have to choose between gearing your wedges to the fairways you normally play or the sand traps, go with the wedge that works best out of the sand. If your course has soft fairways and firm traps, for instance, use a wedge with a little less bounce because that's better in the trap. Of course trying to find the compromise between the two wedges is the best idea, but that's never easy. There just isn't one perfect wedge out there.

Don't be afraid to carry two wedges of similar lofts but with different bounces, because there might be specific shots on your home course that require a very specific wedge. If you have a 56-degree wedge with little bounce to play out of your firm sand traps, for example, it's okay to carry a 58-degree wedge with more bounce for that pitch shot you always find yourself playing from the tall and soft rough next to the sixth green.

I like to see the average player gap their wedges by 4 degrees. So if his pitching wedge is 48 degrees, he might go with a 52-degree gap wedge, a 56-degree sand wedge, and a 60-degree lob wedge. Tour players can get away with 2-degree gaps in their wedges, because their high level of skill lets them play a wider range of shots with each wedge.

There's a tremendous amount of give and take with the Tour players. Sometimes they might change their technique, so they want a wedge with a completely different bounce and loft combination.

A Beauty Returns: Wilson Dyna-Power '58 Wedges

Since Gene Sarazan designed golf's first sand wedge for Wilson in 1932, the company's wedges have been used in more than 1500 PGA Tour wins. Now Wilson has developed the company's classic Dyna-Power '58 clubhead design into a family of five new wedges. Wilson's newer patented Fat Shaft adds stability to these clubs, which also feature their original Fluid Feel shaft to sole hosel construction. Bob Mandralla, who designed irons and wedges at Wilson for everyone from Arnold Palmer to Nick Faldo, produced these beauties.

Shafts: Start Your Engines

The shaft is often called "the engine of the golf club," although it has traditionally taken a back seat to the more visible design, technology, and brand name status of the clubhead itself. Today's equipment landscape, however, reveals a very different scene, with graphite shafts for drivers quickly gaining a celebrity status and an autonomy all their own. Of course, along with the marquee billing comes the bill, and shafts such as the Fujikura Speeder 757 (and others), which some of the game's top pros, including Phil Mickelson, have played, retail for well over $200. And that's without a clubhead attached to them!

In fact, when golfers walk into a quality club-repair shop, the first thing they often see are 15 high-end shafts lined up against the counter. That's because golfers in growing numbers are choosing to reshaft their drivers rather than buy new ones when they are not happy with their clubs' performance.

Even so, important questions remain. What makes these shafts so expensive, and are they worth the money? What level of player do they help most? How does a golfer choose the shaft best suited for his or her game?

Most of the new shafts' high price tag originates in the extremely strong and superior NASA-grade graphite, which allows shaft companies to manufacture high-performing shafts as light as 50 grams each. (A standard steel shaft, by comparison, weighs about 120 grams.) What's more, companies make each model of these custom shafts in relatively small quantities, and this labor-intensive enterprise requires diligent quality control, which also raises their prices.

Robb Schikner, Vice President of research and development for Graphite Design, on shafts today and tomorrow

The biggest advance in shafts today is definitely in their materials. Just 10 years ago, the fibers and resin systems used in making graphite shafts were far less durable compared to the fibers and resin systems you can get today, so you don't have the same rate of shaft breakage now like you did then. Also, the manufacturing process of working with the raw materials has advanced a lot today, which also leads to more consistent shafts. Again, 10 years ago it was difficult to manufacture consistent shafts one after another because every batch of raw material with which you started was very different.

I think in the near future, graphite shafts will continue to get lighter. I also think that more of the club manufacturers will start offering custom shafts as options to their consumers in their stock line of drivers. There's been a tremendous growth on the custom side, meaning golfers are ordering better shafts directly from the manufacturer, sometimes at considerable additional cost. This may encourage companies to start putting the higher-quality shaft into their stock drivers right off the bat, so golfers can buy them off the rack in the golf shop. The cost will be more to the companies, which they'll have to pass on to the golfer, but I think golfers will be willing to pay a little more for a better shaft because it makes a positive difference in their games.

Graphite Design's
Tour GAT 75

However, better shafts need stronger and higher-quality graphite. One of today's 400cc titanium driver heads puts more stress on the tip end of a shaft than its smaller-sized predecessors. This naturally results in more twisting, or torque, at impact. The better material of the new shafts means that they have less torque (meaning they twist less) than both graphite shafts of years past and their contemporary stock shaft counterparts.

Graphite shafts have also improved thanks to better manufacturing processes and quality control. Companies no longer grind away a lot of the excess graphite used to create their shafts. Excessive grinding (still done on less-expensive shafts) compromises a shaft's structural integrity and the performance of the shaft.

Most shaft companies make their graphite shafts by layering sheets of graphite to a desired thickness and distribution pattern around a mandrel. This allows them to engineer in the exact stiffness, flex points, and degrees of torque that they want. The inherent pliability of the graphite also allows shaft makers great latitude in their designs, which results in shafts with a wide range of performance characteristics. Shaft maker Aldila has developed the Aldila One intracomponent shaft, with its fascinating mixing and matching capabilities. For example, the company can make one version of the shaft with a stiff upper portion for stability and a flexible tip or lower end for a softer feel and a high trajectory. They can then take the exact same stiff top part and attach it to a different and firmer bottom section for a shaft that still feels firm but launches the ball at a lower angle. The NV, also from Aldila, features a microlaminate technology that combines very thin sheets of graphite layered in several different patterns in one shaft.

Benoit Vincent, Chief Technical Officer of club and ball research and development at TaylorMade-adidas Golf, sees this growing plethora of shaft options as a good thing. He points out that a number of golfers' performance needs fall in between a clubmaker's stock shaft offerings, and that such golfers can certainly benefit from custom shafts.

Clearly the golf pros, with their high swing speeds, get the most out of these high-strength, low-toque shafts. However, more and more experts agree that these better shafts help the higher-handicapper as well. The less-skilled player miss-hits shots more often than the pros, and because the lower torque of a better shaft keeps the clubface squarer at impact, this results in straighter shots.

Nevertheless, golfers continue to express mixed opinions about these shafts, ranging from skeptics who do not believe they can help anyone to fanatical devotees who claim they have added 25 additional yards (or more) to their drives. The key for the golfer is matching the right shaft to his or her own swing, and this is something a golfer's pro or club-repair person can best help them do.

Aldila's NV

TechTip

Shaft Fitting and Refitting

Like a great supporting actor, the role of the shaft on a golf club is to get out of the way in order to let the clubhead do its work. From this perspective, the perfect shaft would form a theoretically weightless connection between the grip and clubhead, so that you could directly control the clubhead with your hands during the swing. Although this, like any fantasy of perfection, is impossible, if not ridiculous, to approximate such a blissful union you need to follow these two fitting parameters when selecting a graphite shaft:

1. Choose the flex of the shaft based on the speed of your swing.

   Only players with swing speeds of 95 mph or more should choose stiff shafts. Other players should use regular flex shafts, or, for slower swingers, more flexible shafts yet, which are sometimes designated as A, Ladies, or Senior flexes.

   Shafts that are too flexible usually result in shots that both hook or pull and fly too high. Excessively stiff shafts generally produce slices or pushed shots and a low trajectory. Golfers should choose shafts that they can hit straight and high (but not too high) on a consistent basis.

2. With the driver, the length of the shaft is an important consideration.

For years, the rule here has been that golfers should swing the longest shaft they are able to control. Until now, 45 inches served as a kind of cut-off marker beyond which the average golfer shouldn't venture. However, as clubheads exceed the 400cc mark in total volume, shafts have lengthened about an inch as well for a couple of reasons.

First, there's perception: A clubhead that large attached to the end of even a 45-inch shaft moves from mammoth to monstrous. Longer shafts restore a comfortable sense of proportion. Second, the larger clubheads give golfers more surface area on which to miss-hit their shots without sacrificing distance or accuracy. A longer shaft creates a longer swing arc and speed, which translates into longer shots. In other words, today's graphite shafts free golfers to swing aggressively at the ball, while minimizing their worries about hitting their drives too far off-line. The point to remember here is that even though drivers have longer shafts than in years past, they are not harder to control because the drivers' larger and more forgiving titanium clubheads.

Putters: Rolling, Rolling, Rolling

Jack Nicklaus once said that a golfer would "stand on his head" if it would help him or her putt better. Indeed, putting and putters represent the most creative and personal, if not playful and quirky, expressions of golf technique and club design.

In fact, the entire modern era in golf equipment arguably started when Karsten Solheim, an engineer for General Electric, took up the game of golf at age 40. Frustrated that it took him 10 strokes to get the ball on the green the first time he played, Solheim realized that the fastest way to lower his score would be to improve his putting. So he built a putter for himself in his garage based on the design of a tennis racket, with its wood frame supporting its strings.

Of course, instead of wood, Solheim placed some extra metal on the heel and toe of a prototype putter. In so doing, he built the first perimeter-weight golf club. He then started a company called PING, which fueled a burgeoning golf industry obsessed with a passion to improve weight distribution in golf clubs that hasn't died yet.

In fact, one of PING's newest putters, the JAS, which features a titanium body with pure tungsten weights positioned on its heel and toe, represents the company's most extremely perimeter-weighted putter to date.

Perimeter-weighted putters from PING

In the mid-1990s, a then little-known entrepreneurial clubmaker called Odyssey single-handedly popularized face inserts in putters. Odyssey's young chief engineer and club designer, Brian Pond, developed an extremely lightweight synthetic polymer insert called Stronomic (a name Odyssey just made up), which replaced the far heavier steel as the club's putterface.

Not only did this soft material feel great when stroking putts, it also allowed Pond to take this saved weight and reposition it around the clubface's perimeter for greater stability. Subsequently, face inserts comprised of various recipes of polymer, lightweight metals, and/or metal alloys (such as aluminum, Wolfram, Beryllium copper, Cyanamet, and others) have proliferated in putterfaces from a wide range of manufacturers. Pond and several of Odyssey's original executives (including Vikash Sanyal and Brad Adams) went on to start a new putter company called Never Compromise that specialized in expanding the range and properties of polymer putterface inserts.

Callaway Golf purchased Odyssey in 1997 and, almost five years later, broke new ground in putter design with its remarkably successful White Hot 2-Ball putter. In 2003, Cleveland Golf purchased Never Compromise.

An Odyssey putter with Stronomic insert

The original mallet-style club (several iterations of the 2-Ball followed) featured two concentric white discs that are the exact diameters of golf balls mounted on a flange that extends back from the clubface. Golfers find it extremely easy to align these discs with their actual golf ball straight to the hole or to the intended line of their put. Almost as a bonus, golfers have found that the back-weighted clubhead results in putts that roll very smoothly and skid very little (the ideal combination for a putter) off of the clubface.

The Odyssey White Hot 2-Ball putter

The 2-Ball has spawned a litter of large, rear-weighted mallet-style putters, such as Titleist's Futura by Scotty Cameron (used by Phil Mickelson; see sidebar), Nike's Blue Chip Oz, TaylorMade-adidas's Rossa Manza, and Ben Hogan Golf's Big Ben, wielded by Jim Furyk to win the 2003 U.S. Open. (Companies often add tungsten or other heavy metal plugs in the back of the putter head for this extra weight.)

TaylorMade-adidas's Rossa Manza designers evidently understand how important it is that a putter fit a golfer's size, stoke type, and sense of aesthetics—that is to say, his or her eye. To better understand just what makes a great putter, and to build the often one-of-a-kind putters for Tour pros, the company uses a highly sophisticated digitalized putter fitting system. Utilizing cameras and motion sensors, this machine records and measures the golfer, and then creates a virtual 3D computer-generated animated image of the golfer stroking a putt.

Rossa's designers analyze data, including the path of the putterhead, the angle at which the putter strikes the ball, and the degree of loft of the clubface at impact, and then build a putter precisely calibrated to fit each pro's needs. Rossa then uses the information it garners from working with the pros to extend its consumer line, which currently offers close to 30 different putter models.

PING, always an industry leader in clubfitting, now offers online putter fitting with its Specify model putters. After logging on to PING's web site (www.pinggolf.com) and answering a series of fitting questions (about physical size, style of stroke, preference in clubhead design, and so on), golfers can build their own putter by picking and choosing from several components, including head shape and weight, hosel design, length, and lie. They also can go through this process with the help of a trained PING fitter, who must also be a golf pro, at their local course or golf shop. PING also recently introduced its

G2I line, which takes the best historic designs from the company's more than 300 models of putters and upgrades them with soft urethane inserts as well as minor perimeter-weighting adjustments.

The well-respected teaching pro Todd Sones developed a unique fitting procedure for his line of putters, which were developed for Tommy Armour Golf. Called the Tri-Measure Fitting System, the system has a fitter place the golfer in his or her normal putting setup position. He or she then measures the length from the player's lead wrist crease straight down to

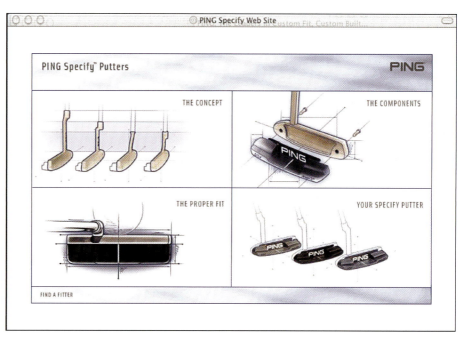

From PING Specify Putters' web site

TechTip

Scotty Cameron has an extensive and fascinating web site, www.scottycameron.com, where he discusses in an in-depth yet entertaining and accessible manner, the technology, design philosophy, and physics of both his putters and of putters in general. He also guides viewers on a tour of his high-tech Putter Studio.

the ground, and then takes another measurement from that point across to the inside edge of the ball. By using the Pythagorean theorem, the fitter can determine the proper length of the putter, which the company custom builds using a putterhead weight that matches the shaft length.

The birth of the belly putter, first used by Paul Azinger in 1999, is not only worth noting because of its significant use on Tour (by some of the game's best players, including Fred Couples, Colin Montgomery, and Vijay Singh, among others), but also because it further illustrates Jack Nicklaus's insight into the unselfconscious measures golfers will take to sink a putt.

Golfers essentially stab or anchor the cap end of this (approximately) 41-inch club right into their navel, which establishes an extremely stable pivot point for the putting stroke. After taking a normal putting grip on the extended handle of the club, golfers can swing the putter back and through the ball almost effortlessly; there is very little wobble or erratic hand

Broom handle putter

motion because of the belly anchor point. Golfers can feast on a full plate of belly putter info from the belly putter web site: www.bellyputter.com.

The belly putter grew out of, or, one might say, shrunk down from, the so-called long or broom handle putter. Golfers stabilized this club by pressing its top end snuggly against their chests, so that it, like the belly putter, swings like a pendulum back and forth through the ball. All the major putter manufacturers make both long and belly putters of the most popular putterheads, which they add weight to in order to accommodate their longer and heavier shafts.

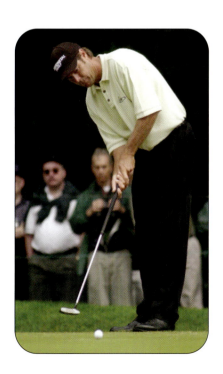

Belly Putter

TechTip

The www.bellyputter.com web site explores the facts and growing folklore surrounding this club.

Phil Mickelson, PGA Tour player, on why he uses Titleist's Futura putter by Scotty Cameron

Phil Mickelson

I started using the Futura in October [2003], and I've seen the difference in the roll of the golf ball of this putter's face on Scotty Cameron's camera in his Putter Studio. That's why I stick with it: It has to do with the back weighting and the angle of ascent at which the face starts the ball rolling. In other words, the less loft of the putter makes the ball roll very easily and quickly off the face, and I feel like I've become a better putter because of this. Now it has taken a little bit of getting used to...it's not the most aesthetically pleasing thing to look down at...in fact, it's downright ugly. But once I have gotten used to how it looks, I feel like I putt better with it.

The Frankly Putter, from Frank Thomas (www.franklygolf.com)

The Frankly Putter

Equipment expert Frank Thomas' new Frankly Putter takes you back to the future. The club presents a classic brass "Bulls-Eye-shaped" putter head with a host of hi-tech amendments and amenities. Known as the inventor of the graphite shaft (in 1969) as well as for his 26 years as USGA's Technical Director (where he ruled on golf clubs' "legality"), the erudite Thomas integrates the best of what he's observed in putter design into the Frankly.

This visually striking putter blends several ideas that reflect, in Thomas' own words, an "understanding of what a golfer needs." First, an innovative aiming guidance line extends down the shaft and across the top of the clubhead. Next a durable, translucent polymer coating encapsulates the flanged club head, provides a soft feel at impact, and visually reveals the club's heel-toe tungsten cylinder perimeter-weighting system. As for the shaft, it's graphite of course, which, as Thomas says, "adds more feel and improves clubhead balance."

"Most pieces of equipment and machinery, other than golf clubs, come with users' guides," says Thomas. "We not only include a printed *Putters Guide* that covers the mechanical and 'mental' fundamentals of putting, we take it a step further by providing each registered club owner with more in-depth online instructional support, which we will continue to develop and update with input from top teaching pros and sports psychologists."

Grips: Hold On

There are only three parts to a golf club: the head, the shaft, and the grip. Considering that golfers only come in contact with the grip during their swings, it remains a mystery why it hasn't received more attention.

A discussion of grips must include the grip's size and weight, its friction or "tackiness" (meaning how well it facilitates a secure hold on the club), its durability, and last but not least, the grip's feel, or tactile qualities. Grip makers today have expanded their catalog of products to include more options in each of these categories than ever before.

Bob Lamkin: President of Lamkin Grips, on the technology of grips

The grip has really evolved in pace with the technology of equipment in general. Grips have traditionally been manufactured predominantly from natural rubber, and with natural rubber, you get a lot of inconsistencies, because you are dealing with Mother Nature. Over the last several years, we've moved to synthetic rubber or plastic polymers in making grips, which are much more consistent and more durable than rubber. What we've really been able to achieve with these synthetic materials is a consistency in the manufacturing of the grips. For example, if a clubmaker comes to us and says, "We want Lamkin to produce a grip that is 4 grams lighter, or 4 grams heavier," whatever it might be, we're able to achieve the weight variances in that grip because of the consistency in the material and the vastly improved manufacturing processes we now use. In short, grips are much better now than they have ever been.

Let me give you another example. Not long ago, if we wanted to make grips lighter we had to do that by adjusting the weight of the grip's core size, whereas today we can keep the grips same dimensional size and adjust their weight through the materials we use to make them. It's a huge advantage in terms of ease of manufacturing.

Looking into the future of grips, I'll say that we're going to keep pushing new materials, such as the ones we're experimenting with right now, that are not endemic to golf grip manufacturing. I'm very excited about these materials, because we'll be further able to offer a wider range of degrees of hardness or softness in the grip, different weights, different colors, and upgrades in playability and function, as well.

Companies today are able to blend and adjust the polymers or synthetic rubber compounds they use in manufacturing grips into a wide range of weights. Whereas a standard grip weighs 51 grams, lighter grips, such as the Winn Grips, weigh 42 grams. Golf Pride's new Decade Lite Cord grip measures only 38 grams, whereas others from different manufacturers go as low as 35 grams. Regripping with a lighter or heavier grip significantly alters a club's overall weight and feel, so golfers have to take a couple of things into consideration when they do (see sidebar). Most models of grips come in a range of diameter sizes or thickness to fit small, medium, and large hands.

Grips achieve their friction, or the degree of adhesion they establish with the golfer's hands, in several ways. Golf Pride, Lamkin, and others have, for years, manufactured rubber compound grips laced with cords that provide a secure hold on the club even in rainy weather. Lamkin's new G3 Quarter Cord grip, for example, features a small and separate heavily corded area for extra friction where the glove hand (the left, for right-handed golfers) holds the club. This is where the most friction occurs during the swing. The lower non-gloved section of the grip features a soft blend of cork and natural rubber.

Lamkin's G3 Quarter Cord grip

Other styles of grips offer perforated surfaces configured into different designs, whose roughed-up edges add friction and traction. Several wrap-style grips (such as Golf Pride's Tour Wrap and Lamkin's Perma Wrap) have pronounced ridges that circle diagonally down the grip, creating raised seams that provide a secure holding surface on the club. Water beads up on these wraps and sits on the surface of the grips, which, when wiped off by the golfer, retain all of their innate tackiness. Many golfers prefer grips with a reminder, or a thin extra ridge of rubber molded into the underside of the grip, which helps them place their hands on the club in a consistent manner.

Lamkin's Perma Wrap

Often a grip's friction-producing strategy overlaps with its tactile or feel qualities. For example, Golf Pride's Tour Velvet grip and Lamkin's Crossline grip feature closely grouped and small perforations. By puncturing the grip's surface, these tiny holes soften the grip and provide a softer, more comfortable sensation that many golfers enjoy. Other companies, such as Winn, Avon, tacki-mac, and Karakal, specialize in even softer grips. (Golf Pride offers such a grip as well.) These products also absorb the shock of impact very well (oversized versions of these grips provide even more cushion effect) and appeal to golfers with tender, injured, or arthritic hands and joints. The downside of these grips, however, is that they have a tendency to slide or become marginally unstable at impact, so stronger players generally avoid using them. To their credit, however, most of these companies have developed somewhat firmer versions of these very soft grips for golfers with higher swing speeds.

TechTip

Mike Chwasky, Senior Editor, *Golf Tips Magazine*, with some tips about grips

Golfers need to be aware that if they change the weight of their grips they can significantly change the feel of their clubs and their performance as well. Before they regrip their clubs, they should know the weight of their current grips, and realize if they regrip with lighter grips they will increase the swing weight (which is the relation between a club's grip and head weights) of their clubs and make them feel heavier. If they regrip with heavier grips, their clubs will feel lighter. In addition to a lighter grip making the clubhead feel heavier, it will also give the shaft a little more flex. Vice versa, the heavier grip/lighter swing weight combination stiffens the shaft marginally. So if a golfer likes the way his or her clubs feel and it's time to regrip them, he or she should choose new grips that weigh the same as the old ones.

Also, when golfers reshaft clubs with lighter shafts, which many players are starting to do, they not only decrease the static weight of the clubs, they also decrease their swing weights.

However, by putting a lighter weight grip on the reshafted clubs, they can balance that equation out and still maintain the original clubs' swing weights.

Understanding how a grip's size affects the swing is very important. The need for golfers to find the proper grip size is greatly underrated. Grips that are too big tend to inhibit a good release of the club, which decreases power and clubhead speed and tends to promote slicing. Conversely, grips that are too small tend to activate the hands more and promote hooking and erratic shot patterns.

Most recreational players should consider changing grips once a year, but how long a grip lasts is, of course, directly related to how much a golfer plays and practices. Golfers can, however, significantly prolong the lives of their grips by cleaning them with warm water and soap fairly regularly. The pros, for example, have their grips cleaned, usually by their caddies, after every round or practice session.

Women's Clubs: Ms. Hits

Annika Sorenstam

Photo Copyright Paul Lester

When discussing technology and women's golf clubs, one must keep in mind the wide range of skill levels, swing speeds, and body types that comprise the spectrum of women players. At one end of this spectrum, Annika Sorenstam drives the golf ball more than 250 yards (not to mention teenage prodigy Michelle Wie, who hits it considerably farther). College golfers may hit it 200 yards, whereas athletic recreational players may average 180 yards off the tee. Toward the other end of the spectrum are older women, beginners, and less-athletic players, who may hit the ball 150 yards or less.

All these players have different needs. However, the same equipment issues apply to women as to men—after all, the golf ball doesn't know who hit it. Areas of concern include how to maximize COR, choosing the right loft and shaft for a driver, and finding what golf ball best complements a given player's club and swing profile.

Virtually all the LPGA Tour pros and top women amateurs play with what, for a lack of another name, must be called "men's clubs." There are, however, plenty of strong women amateur golfers who might not be strong enough to play the heavier, stiffer shafted men's products, but are a bit too powerful to play clubs meant for the 150-yard hitter. These players can pick from the high-end clubs offered from several companies specializing in women's clubs, which are generally shorter and lighter than men's clubs. The advantages include a list of technologies that will specifically improve a woman's game: shafts with low bend or kick points; clubheads with lower centers of gravity to help get the ball well up in the air; lightweight and smaller grips for comfort and faster swing speeds; a choice of standard or offset heads in both higher-lofted drivers (11 degrees and higher) and higher-lofted fairway woods (including 7, 9, 11, 13, even 15 fairway models) that achieve maximum height and distance even at slower swing speeds.

Betsy King

Photo Copyright Paul Lester

Top companies that offer the best line of women's clubs include Cobra, Callaway, Square Two, LJC Golf, Nancy Lopez Golf, Lange Golf, Cleveland Golf, TaylorMade-adidas, Nike, PING, and others.

For women who drive the ball 150 yards or less, the solution is not just a suitable driver: Equipment makers are beginning to pay more attention to the entire set of clubs for these women. Here's why. Generally, a woman who drives the ball 120 yards will hit her pitching wedge about 80 yards, or just 40 yards shorter than her drive. That 40-yard difference represents, in golf lingo, the player's "distance gap." (A strong male player might hit his driver 250 yards and his pitching wedge 120 yards, for a 130-yard gap.)

One reason for the size of this gap is that this player cannot generate a significant amount of spin with any of her clubs. If she hits her driver with 4000 rpms (revolutions per minute), she hits her pitching wedge with 5000 rpms. The spin rate is almost constant through the set; therefore she hits the ball almost the same distance with many of her clubs. This compressed distance gap, as Benoit Vincent of TaylorMade-adidas Golf calls it, represents the heart of the problem for these players, and it's one that the clubmakers have just begun to recognize. The solution, according to Vincent, is not to try to build drivers that hit the ball farther but to develop a set of clubs that as best as possible fills the yardage gaps from the driver to the pitching wedge so that as women get closer to the pin, they have clubs that get them to the green.

To determine the club designs and specifications of such a set, TaylorMade, Callaway, and other companies have started to put women on their launch monitors to record their impact data (including ball speed after impact, spin rates, launch angles, and so forth), which they use to quantify this compressed distance gap. From this information, they set out to design a different kind of set for these women with very slow swing speeds.

The strategy that works best in addressing this 40-yard gap involves creating slightly larger gaps between the lofts of each club in the set and minimizing the number of clubs this kind of player would carry. This spaces out the distances between shots a bit more evenly. A set of clubs designed specifically for slow swing speeds might consist of a 14-degree offset driver; an 18-degree 3 wood; a series of shallow-faced fairway or utility woods with lofts of 20, 24, and 28 degrees, respectively; a 7 iron of 32 degrees, an 8 at 38 degrees, a 44 degree 9 iron; a pitching wedge of 50 degrees; and a 56-degree sand wedge. The total number of clubs, including a putter, is only 11, when the rules allow golfers to carry 14, but this may be all that this type of golfer needs to play her best. TaylorMade-adidas's new Miscela clubs for women represents such a set, as do Callaway's GES (Game Enjoyment System) and PING's G2L. These companies are also bringing out clubs designed for very slow swing speeds, and other companies will no doubt follow soon.

TaylorMade's Miscela irons for women

Debbie Steinbach, Former LPGA Tour player and currently an LPGA Teaching professional, author of *Venus on the Fairways*, on women's clubs

Debbie Steinbach
Photo Copyright
Arlene Spring

I think it's about time that manufacturers realize that women represent a big segment of their market. So it's good that they're finally focusing more attention on the needs of women golfers and going beyond the cosmetics of the golf clubs. Having said that, I don't believe that you can completely separate men's and women's clubs, but have to think about fitting clubs to the needs of each individual. Ultimately, companies need to be making clubs for individuals. The fit is the most important element, not whether it says "man" or "woman" on the shaft or the grip.

Whether you're male or female, you have to go out and hit clubs first before you buy them. Just don't buy the ones that might look good to you on the rack in the pro shop. I will say this about women new to golf: I don't think they need to buy a full set of clubs. When they're just starting in this game, why make it so complex? Most women don't take a rocket scientist's approach to golf, anyway: They play the game for recreation, and don't want to dissect it. It's a wonderful solution to the difficulty of the game for a woman to start with a smaller set of seven clubs so that she can see whether she really will get hooked on golf before making a larger investment.

Debbie Steinbach can be reached at venus@venusgolf.com.

Junior Clubs: Spring Swings

Historically, equipment makers building clubs for kids simply found the lightest men's shafts available, cut them down, and then attached them to standard iron and wood heads. In short, they never seriously addressed the issues of the golf club as an integrated whole for this group of players. Even major companies used to use the same heads for kids as they did in their standard clubs. As a result, their junior products were both too heavy and too hard to swing. However, proliferation of quality lightweight shafts in the early 1990s greatly improved the performance of both women's and juniors' clubs, and encouraged equipment companies to think even more progressively about their junior line of products.

Companies such as U.S. Kids (www.uskidsgolf.com) and LJC Golf (www.lajollagolfclub.com) understand that juniors needed clubs equipped not only with lightweight and shorter shafts, but with lightweight clubheads as well. Rather than a junior driver head weighing 220 grams (as in years past), they now fall into the 150 to 170-gram category.

LJC's clubs for the youngest players

Lighter-weighted iron heads feature forgiving cavity-back shallow-faced head designs, with low centers of gravity to help kids hit the ball higher.

LJC Golf does a great job of matching a shaft's flex to a kid's strength and swing speed. This allows the golfer to feel the weight of the clubhead throughout the swing, which facilitates the kind of smooth and rhythmic tempo essential for playing good golf. The company has patented graphite shafts for juniors, and they have made sure that even the shafts in sets for the younger kids have sufficient flex.

Kids' clubmakers now also manufacture small golf bags, gloves, golf shoes, hats, shirts, pull carts, and other accessories so that juniors can feel like "real" golfers and not just tag-alongs.

If the clubfitting process has become an accepted and ubiquitous part of the golf club buying experience for adult players, it has become, possibly, a more important part of buying clubs for kids. Instead of parents having to purchase the shortest and lightest sets of adult clubs available for their kids (legends such as Arnold Palmer and others started playing golf with women's clubs), or cut down their own old clubs with the idea that their kids "will grow into them," U.S. Kids, along with LJC Golf and others, have developed color-coded clubfitting programs that ensure that kids play with clubs that fit them "here and now." LJC Golf's sets become slightly heavier, with marginally stiffer shafts, thicker grips, and more upright lie angles as the kids they're intended for grow bigger and stronger.

Sets for kids age 3 to 5 or 6—such as U.S. Kid's Red System, and LJC's Snoopy set, with Charles Schulz's classic characters colorfully painted onto the clubheads—make the old expression that "golf is a game for a lifetime" truer than ever before.

Paul Herber, President of LJC Golf (www.lajollaclub.com), on junior clubs

Age 12 is the tricky or funny age for kids and golf clubs. At that age, a kid can play like a young adult or like a little kid, depending on his or her level of physical maturity. At LJC Golf, we design a teen set for the 12, 13, even 14 year old who is not that strong yet. Those that are can start playing with adult lines of clubs. All kids that age, it seems, want to play adult clubs so badly. If they do before they're not strong enough, the clubs will be too heavy for them, which can lead to some bad swing habits.

In my opinion, the shaft is the key to junior clubs. That's why we offer 18 different shafts for kids of all ages, sizes, and strengths, all of which have the appropriate amount of flex built in to them. Companies that just cut down shafts for the smaller kids are actually making shafts that are way too stiff, which hurts kids more than it helps.

It's a good idea for kids to try the clubs they are going to play with first, if they can find courses that have demo clubs. I also recommend to parents that they test the junior clubs themselves before buying them for their kids. Each year we have a tournament for adults where they play with sets of our junior clubs. The best players have to play with our Snoopy metalwood and 7 iron that have been designed for 3 to 6 year olds. It's amazing how far they can hit the ball with these clubs. Hitting a quality junior club for themselves might enlighten parents about their quality.

The real problem we face isn't convincing parents to play with, but getting them to pay for, the very real technology we build in to our junior clubs. It's not that parents don't care, they just don't understand the technology well enough. They know they want to start their kids in golf, and if they get them a set of clubs that doesn't break, they're happy. But for kids to enjoy golf now and throughout their lives, they have to find it fun, and playing with quality equipment along with good lessons are two important first steps in that journey.

Senior Clubs: Autumn Rhythms

Any discussion about "senior" men today seems shrouded in a kind of sensitive ambiguity: On one hand, society used to designate 65 as the senior signifier, but then the Senior PGA Tour came along where men 50 and older could qualify. As the baby boom generation ages, 50 seems more like a wrinkle away from Woodstock than it does from the grave. Even the Senior Tour changed its name this year to the Champions Tour to try to take the edge off of the aging process and to refocus its attention on the life and golf experiences of its players.

Nor is it easy to concretize that moment when a male golfer crosses the rubric from a "regular" to a senior player. A lot of senior men have been playing golf for many years and have grooved their swings and technique to the best of their abilities. Next, an increased emphasis on golf-specific stretching and strengthening programs has many seniors in better shape at 55 or 60 than they were at 35 or 40. Finally, advances in club and ball technology have truly given back what nature has taken away, and pros and amateur players in their vintage years express with almost intoxicated glee the fact that they are hitting the golf ball farther now than ever before.

For these reasons, senior golf clubs reveal a gradual, if not seamless, transition from the technological standpoint. Perhaps adding a little loft to slightly lighter drivers and fairway woods, and switching to little weaker (that is, more flexible) shafts are the basic steps senior men might take to keep pounding the ball out there with their young "flat-belly" counterparts (grandkids?).

Jack Nicklaus

Photo Copyright
Paul Lester

Companies such as Cobra Golf, a long-time industry leader in clubs for the senior male player, now offer 11- and 12-degree drivers, and offset models as well, which give an extra little lift or height to shots from the tee. Rather than struggle with that strong 14-degree 3 wood off the fairway, seniors are turning to the higher lofted 5, 7, even 9 woods. The legion of new utility or hybrid-type fairway wood/iron couldn't find a more appreciative or fitting audience than the senior golfer.

The same is true for today's lineup of putters. As the back stiffens a bit more quickly during a round and bending down to get the ball out of the hole becomes as much of an experience as it is an event, seniors are turning to the long and belly putters, which are palliative for both situations. Again, it's a lucky coincidence that these products have come into their own just as a host of boomers are "making the turn." Maybe golf is a game for old men, as the old cliché had it all along.

Cobra Golf's SS Senior Driver

Jeff Harmet, General Manager, Cobra Golf, on club technology for seniors

Many of the advances in golf clubs that we have seen in the general market obviously carry over into the seniors. The most obvious are thinner and more lively clubfaces on drivers for added distance, and larger clubheads that are more forgiving. Even with these oversized drivers and clubheads, manufacturers such as Cobra Golf have figured out how to redistribute some of the clubhead's weight lower, farther back, and toward the heel of the clubs, adding height to the shot, countering the higher-handicapper's tendency to slice, and promoting a nice controlled draw.

At Cobra we realized long ago that seniors need a different kind of product relative to the other markets out there, such as the avid male golfer and the woman golfer. Some of the things we did then, with our Cobra SS Senior driver, were to build it with an offset clubhead and 11 degrees of loft. Typically, senior golfers have slower swing speeds than avid young male golfers, and these design features assist in getting the ball airborne for longer drivers. They also promote a left-to-right draw, which is another way of saying they are anti-slice.

The swing weight on our senior model is a little lighter, in the C9 range, again to promote greater carry distance. Our senior driver has a softer or more flexible shaft, and a lighter 50-gram shaft, whereas some of our avid male driver shafts are 65 or 75 grams or higher. We also have a longer shaft in the driver, 45.75 inches versus 45 (25 in the regular model), again to promote a faster clubhead speed for more distance. Finally, we use slightly softer grips as well, both for comfort and for shock-absorption reasons for seniors who may have arthritis or other health issues.

We've lightened the swing weight for the seniors as well, and we've weakened the lofts on the 7 iron down to promote greater carry distance and to get the ball up in the air. Finally, we've widened the sole of our senior irons, to get the center of gravity down and farther back to promote greater carry distance.

In a way, you hate to call these senior clubs. The reality is that these clubs were designed for people with slower swing speeds. Ultimately, for anyone, it comes down to trying different products to see which will produce the best distance, accuracy, and trajectory for each player.

Clubfitting: Fit for a Swing

Imagine a track star suiting up for the 100-yard dash. It's time to lace up his running shoes in the locker room, so he goes over to the shoe rack and grabs the first pair he sees, even though they are three sizes too small. No one could run their fastest in such shoes, yet every day, scores of golfers peruse the club racks of golf shops and lay down as much as $1200 (sometimes more!) for a set of irons, and an additional grand for a driver, fairway wood, and putter that, although built to standard specifications, don't fit them at all. Help, however, is here, as virtually every major company today offers nationwide clubfitting programs that all but ensure that golfers will buy clubs well fit to their size, strength, and, most importantly, individual golf swings.

A clubfitting session can be quite simple, with a trained fitter, who is often also a teaching golf professional, supervising the golfer as he or she hits off of a plastic lie board for iron fitting or tests a series of drivers for the one that fits best. Then there is MacGregor Golf's deluxe version, which blends traditional fitting and the kind of fawned-over customization generally reserved for the clubmaker's very best Tour players.

For about $8000, MacGregor Golf will fly golfers to their corporate headquarters in Albany, Georgia, where Don White and Dave Wood, two legendary club designers, work with them in building *from scratch* their set of dream V-Foil Personal forged irons. Golfers actually work with CAD technology to produce their one-of-a-kind set of irons, which includes the clubs' head shape and size as well as a sole ground to match their individual swings and normal playing conditions.

Burrows Golf's fitting system offers many shaft/clubhead combinations.

A few years ago, PING made a significant investment in their already industry-leading clubfitting program when it introduced its Wrks department (pronounced Works). Customers can visit PING's plant in Phoenix to design their own clubs, and they can order irons, metalwoods, wedges, and putters through PING's Wrks department, with a wide range of detailed and highly personalized specifications. These include special sole grinds, unusual (that is, not standard) lengths and lies, swing and overall weights, grip sizes, and types and custom shafts. Additional costs vary depending on the options ordered.

Henry-Griffitts, like PING, is a true pioneer in the clubfitting category. For example, they were the first company to use a lie board in iron fitting (see sidebar). They also introduced interchangeable shaft fitting, by which shafts of different lengths and flexes snap on to different clubheads during the fitting sessions to help golfers select the best combination for their swing. A new club company, Burrows Golf, employs a similar snap-on, interchangeable, shafts-fitting system to fit golfers for their MAC driver, instead of using stock shafts. Burrows offers several of today's highest-quality graphite shafts, such as Fujikura, Graphite Design, Aldila, Harmon, UST, and others. The company matches these shafts with 21 different metalwood head options, for a total of more than 600 fitting combinations.

The TaylorMade approach to driver fitting reflects their understanding that different swing types require different driver designs. To this end, they offer three distinctly different drivers in their new R500 XD and Tour Preferred series lines: the R510 Tour Preferred at 390cc for very skilled players who want optimum workability (and the option of hitting fades or draws); the R540 XD 400cc driver, which adds spin to the ball for players with slower club-head speeds; and the largest driver, the R580 XD 440cc driver, which produces the highest trajectory with reduced spin. A launch monitor can verify and add confidence to the golfer's choice of one of these clubs.

Other top driver makers, such as Callaway, Cleveland, and Titleist, also offer drivers that fit a golfer's specific skill level, but players must learn the somewhat coded language they use to label these products. For example, Callaway's Great Big Bertha II driver comes in a Pro Series model, which means it has a kind of square-to-slightly-open clubface (as opposed to most oversized drivers that have slightly closed clubfaces to prevent slicing), and an internal weight placement that promotes a straight or fade ball flight (whereas the standard model promotes a draw) that better players traditionally prefer. Cleveland's Launcher 400 driver series doesn't give a name to different clubs meant for different skill levels, but it implies it through its loft designations. Their drivers with more than 9 degrees of loft and higher feature more closed faces and hook-biased, internal, weighting schemes, whereas those lower than 9 degrees have the kind of squarer faces and straight-to-fade weighting biases that the pro or strong amateur often favors.

The introduction of launch monitors in the past couple of years has really allowed driver fitters to hone in on a club that presents the best clubhead size/loft/length and shaft combination for each golfer. Simply put, a launch monitor is a suitcase-sized computerized device whose camera or laser records and analyzes a host of impact data during a fitting session. This data includes launch angle, initial ball speed, and spin rates, from which the launch monitor calculates the distance, direction, and trajectory of each test shot. Callaway Golf uses its own launch monitors at the company's headquarters in Carlsbad, California, to fit golfers for their new ERC Fusion and other drivers. Golfers can schedule fitting sessions on Callaway launch monitors at the company's franchised fitting centers in Boston, Las Vegas, and Indian Wells, California. To date, launch monitors are mostly used to fit players for drivers and sometimes fairway woods, although all the major companies use launch monitors (some of which they

Vector's launch monitor

develop themselves) in the development of all of their new clubs, irons, wedges, and even putters.

Companies such as Focaltron (makers of a launch monitor called the Golf Achiever) and Accusport make retail models of launch monitors that sell for about $5000. More and more golf shops use them to help golfers purchase the best clubs (mostly drivers) for their games.

Clubfitting: Basics

- **Iron fitting**—Irons that are too upright for a given golfer tend to produce shots that fly off-line to the left (for right handed golfers), because the heel of the club grabs on the turf and the toe closes too quickly. Conversely, clubs that are relatively too flat lead to shots lost to the right, as this time the toe grabs the ground first and the heel swings through impact opening the club-face. Well-fit irons display clubs whose soles contact the ball and the club flushly and simultaneously, thus producing straight-flying shots. Fitters generally use a plastic lie board to fit golfers for the proper lie angle. Tape is applied to the sole of the club and shots are struck off the board, after which the fitter reads the marks on the tape to determine whether the iron tested is too flat, upright, or just right.

- **Length (for all clubs)**—A golf club well fitted for length allows the golfer to hit the ball an optimum distance without sacrificing accuracy. Golfers should test clubs of various lengths to find the right length for their size and skill level. The consensus among clubfitters has been that golfers should play the longest clubs that they can swing with control.

- **Shaft flex (for all clubs)**—Again, golfers should use shafts that are flexible enough to hit the ball their optimum distance yet firm enough so that they do not sacrifice control. The best method for testing shafts is trial and error.

- **Putter fitting**—Putters are the most idio-syncratic of all the clubs, and selecting a putter is generally a matter of taste. The best rule to follow with putter fitting, then, is just to use what works, and not to be embarrassed if a putter looks a little funny.

- **Wedge fitting**—Wedges are fit in groups, meaning you want to make sure you are carrying the right number of wedges for the types of shots you will need to hit during your round. The key is having the proper combination of wedges to fill what are called distance gaps. (See the section titled "Wedges: Great Escapes" earlier in this chapter.)

Club Repair

Technological advances in golf equipment have made the people who repair clubs important players in today's golf scene. Whereas in years past their jobs were mainly to fix broken gear, their own new glues and gadgets have increased the range of services they can perform while decreasing the time it takes to do them. The repair shop is really an "if you want it now, you can have it now" kind of place these days. Furthermore, these club-tinkering tricksters know both contemporary equipment and their customers' games inside out and backward, so who better to quickly set a wayward golfer's game straight than them?

Kit Mungo is one of these behind-the-scenes heroes. The proprietor of A Better Club repair shop in Ventura, California, Mungo is an accomplished club designer and manufacturer, a single-digit-handicap player, and a part-time equipment technician who works in the PGA, Champions, and LPGA Tour vans at select tournaments. He offers the following experienced and thoughtful Tour of the repair person's fast-moving, though well-measured milieu.

TechTip

Picture Perfect Club Repair

Golf Club Repair in Pictures, 5th Edition, **by Ralph Malby and Mark Wilson provides procedures, charts, tables, information, and methods—and, of course, plenty of pictures. It is the official golf club repair manual for the PGA of America.**

Kit Mungo, proprietor of A Better Club Repair Shop, on the 5-minute solution

Kit Mungo

The first thing that advances in technology has allowed repair people like me do is to attach the head of a club to the shaft and make sure it stays there. What's really exciting about the new epoxies we use to do this is that they can create a bond between the shaft and clubhead in less than 5 minutes that will withstand an impact with the ball at 125 mph of clubhead speed. Just 7 or 8 years ago, this would have been an overnight affair. When it comes to the Tour players, this speeds up the whole process of their club testing tremendously.

Here's how it works. All the pros have driver heads that they really like to stick with. They also want to experiment with different shafts on that head early in the tournament week. Of course, they don't want to spend a lot of time waiting for a new shaft to dry on their clubhead. They want to take that club and get back on the driving range to practice with it. Mostly, they don't want to lose the swing thoughts they had in their heads when they took the club into the van for

reshafting. As I said, in 5 minutes they're back on the range practicing with that club again. Someone such as Vijay Singh will bring two driver heads with him to the range and change the shaft six times in one day on each head. That's 12 changes of shafts into heads in one day. No wonder they called it "5-minute epoxy."

For the average golfer, these fast-drying epoxies make reshafting clubs very convenient. People don't have to leave their clubs at a repair shop overnight anymore. They walk in and out in 5 minutes with their reshafted driver. It's golf's version of one-stop shopping.

We have another type of new glue I want to talk about, rat glue, which is a heavier epoxy that we squirt inside the clubhead to change the club's weight or weight distribution. Clubs such as Titleist's 973 K driver have little plugs on the bottom of the sole plates that we drill out, squirt the rat glue in, and then fill with a new plug. We also can go into the head through the hosel after we take the head off the shaft.

This rat glue has freed us from carrying around a box of driver heads ranging from 196 to 204 grams in the Tour van. Now can take that 196-gram head and make it the right weight for the shaft for which we're fitting it. The real benefit is that we no longer have to go to a longer version of that lightweight shaft to get the club's weight up as we did in the past, and the golfer doesn't have to adjust to a longer club. This stuff dries in 5 minutes, also, so by the time we get the shaft back in, it's ready to go. From the time we put the shaft back in, again, it's just 5 minutes before the golfer can hit the club on the range.

Repair people have always done a lot of regripping, although the solvents we use have only recently gotten better. We used to work with some very toxic stuff that had "Hazardous Material" written all over the bottle. Today's alcohol-based solvents are basically harmless. They also dry, you guessed it, in less than 5 minutes, compared to the couple of hours they took to dry 6 or 7 years back.

The technology of lie and loft machines has gotten much better, which means we can adjust a player's loft and lie more quickly and accurately and cost effectively. For example, one machine now has three setting gauges and interchangeable fixtures to adjust metalwoods, irons, and putters, whereas in years past we needed three separate machines. On Tour we still prefer to use lie and loft machines that are specific to each type of club, but cost isn't a factor out on Tour.

Club manufacturers now provide the Tour vans with molds of their different drivers and fairway woods. We set the clubs into these two-piece aluminum castings, which secure and protect the clubhead while we bend the loft and face angles to the pros' needs.

Last but not least, I want to talk about these small plastic cylinders we use to shim or retrofit driver heads with graphite shafts, because they have become very helpful. Although club companies build some of their drivers with .370 (inch) or .350 hosels, not many shaft makers make shafts with corresponding tip diameters to fit into them. So we use these 1.5-inch plastic shims to reduce the hosel size to .335—the size of the majority of shafts' tips today—and then insert the shaft into them for a very secure fit. We do this both for the Tour player and in the repair shop for our customers.

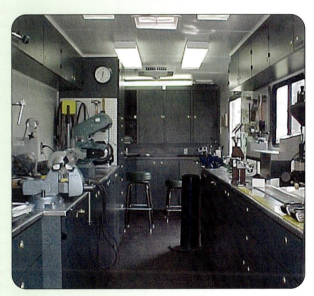

Inside a Tour van's repair shop

Technology and the Tours: Prospectives

The pro game has always been a laboratory for clubmakers, where the best players in the world test new products for manufacturers each year. Although Jack Nicklaus and Jack Nicholson may look like they have the same model of the latest oversized titanium driver in their hands, in fact they each are holding subtlety different clubs.

Because each Jack plays a different game, each needs a different club. Generally, the higher-handicapper (guess which Jack this is?) fights a chronic slice, whereas the Tour star fears the hook. Naturally they need different drivers. Today's 350cc and larger drivers from virtually all the major clubmakers have slightly closed clubfaces, which mitigate the higher-handicappers' tendency to slice. A company can bend or adjust drivers from this closed clubface to a square or slightly open position to meet the Tour players' preferences, but more often they will cast new drivers for them with square or marginally open faces.

Actually, the pros do a lot more work on their irons and wedges than their woods, which mostly involves bending their lie angles and lofts in very specific and individualist ways. Although the lie angles on a theoretically "perfect" iron would all sit flush against the ground (actually, with the toe raised just a bit), and the lofts on such a set would be gapped by 3 degrees, many pros adjust their irons quite differently. Those who want to fade every shot might make their irons flatter than standard, whereas players who prefer a draw would set them more upright. Players who hit the ball high would weaken their lofts, whereas those who hit the ball low would bend them more strongly. Often a pro will strengthen or weaken the loft on one iron to fit one shot on a par 3 hole at a specific tournament course.

Tour players tend to go a little crazy with their wedges. They might grind the leading edge of a wedge very sharply so that they can pick the ball more cleanly off the turf, and then turn around and round off or soften that very edge for more bounce that allows them to hit down into the shot more aggressively. They also may shave a little metal away from a wedge's heel, which allows them to open or close the clubface to hit a wider variety of shots.

Bernhard Langer, World Golf Hall of Fame member, on technology and the PGA Tour

Bernhard Langer

Generally golf equipment has come a very long way in the past few years. The ball goes so much farther and even straighter, and we have so many variations of balls where we can choose from more spin and less spin. The same with golf clubs, and shafts especially. With the drivers, the heads are lighter because they used to be made out of heavy wood. They were solid inside, very heavy therefore. The length of the club was only 43.5 inches; now the players play it anywhere from 45 to up to 50 inches in length because the head is lighter and we can use graphite shafts rather than metal shafts. The lighter material, and larger radius, means more speed and we hit the ball much farther. I probably hit my driver now about 20 yards farther than ever in my life and I should be hitting it shorter, because I'm getting older. Then we have better putters, with more balance, a bigger sweet spot, and more to choose from. The same with all wedges. Now we have wedges with all different degrees from 65 down to 48, anything you want. We also have these utility club 5 woods and 7 woods that you can hit out of the rough, out of fairway bunkers, and out of bad lies. All that has improved the game of golf and therefore the scores are getting better and better.

Also with the grooves on the wedges, we can play square grooves now on loft wedges and sand irons, we get spin no matter what kind of lie we have, whether we are in the rough or semi-rough. The ball doesn't jump or fly anymore; it comes out with spin, it goes a certain distance, and that helps tremendously around the greens as well. And again one of the major things that has improved every professional's game I would think is just the ability to choose the right shaft, the right loft on the driver, the right golf ball to go along with it, and the ability to test it and see what the launch angle is, what the spin rate is, and therefore maximize the launch angle and especially distance and choose from the many variable components that make the perfect club.

In the past, clubmakers and repair people would work on wedges and irons with metal-grinding machines that sparked like the Fourth of July. Today there are also machines for minor touchup work that use a far gentler buffing cloth-like material (one is called Trizact, trademarked by 3M) that polishes the wedge's metal, removes less metal than a grinder, but still adds some effective bounce to the wedge. Titleist's master wedge-maker Bob Vokey talks about fielding phone calls at every hour of the day and night from Tour players to discuss different ideas about grinding their wedges.

Although the general public (as discussed in Clubfitting: Fit for a Swing) has gradually turned to launch monitors for club fitting, Tour pros use them diligently. Perfectionists by nature, these players want to see a certain launch angle, trajectory, and ball flight with their drivers, be it high or low, a draw, a fade, or a straight flight, and the combination of the launch monitor's data and actual testing on the range gives them all the feedback they need. Here are two case studies that show the creative interaction between a launch monitor and Tour players.

According to Benoit Vincent, one of the first things Sergio Garcia did after he switched from Titleist to TaylorMade clubs late in the 2002 season was test new drivers on TaylorMade's launch monitor at the company's Carlsbad headquarters. Garcia has a very steep angle in his downswing, which puts a tremendous amount of spin on the ball, and tends to hit drives that start low then balloon up into the air.

TaylorMade's 580 driver achieves its trajectory via a high launch angle, not by creating high ball spin, so it turned out to be the right club for him at that time.

Chris McGinley, Vice President of golf club marketing for Titleist, says many of the players who use his company's clubs, including Phil Mickelson and Tiger Woods (who played Titleist before switching to Nike during the 2002 season), spend so much time on these machines that they become "launch monitor junkies." According to McGinley, they would often gather around a player hitting with the launch monitor and shout out numbers, trying to guess the launch data before the monitor rendered it digitally.

McGinley goes on to say that a dedicated team of launch monitor operators takes the devices to several Tour events each year, where they work with players on the range either to fit them for new equipment or to adjust the equipment they presently use.

Jeff Colton, Sr., Director of Program Management, who oversees golf club development from inception to finished product for Callaway Golf, recalls Charles Howell III's work with the launch monitor. According to Colton, Howell saw that his launch angle was very low with his rather lofted driver, because he was swinging down on the ball and delofting the club. He actually switched to a *less*-lofted driver, which made him swing on a more level or slightly ascending path through impact. As a result, he raised his launch angle for higher and longer drives.

Amy Alcott, LPGA Hall of Fame member, on technology new and old, and the importance of individuality when selecting equipment

Amy Alcott

Although I believe technology has revolutionized the sport, I'm also a believer that if you're a great swinger of the club, you can play with anything. You can play with a broomstick, for that matter. Mainly, today's lighter shafts and clubs with larger heads and large sweet spots have helped to minimize mistakes. With lighter clubs, you can hit more balls and you can practice harder. This is why LPGA pros, and all the pros and amateurs too for that matter, are going to these fitting centers, which have all of these high-tech diagnostic tools to fit them properly. They're trying to get that edge.

A lot of Tour players have always tried the different technologies in equipment as they came along, but I actually won all of 30 professional tournaments using the same old set of Dunlop blade iron clubs. All I did was change the grips on them 20 or 30 times. As technology was changing and making clubs stronger that hit the ball farther, I stuck with clubs that had weaker lofts for a long time, even though I couldn't hit the ball as far with them. But I believed in them. I knew I could play with them. They were like my best friends, and I knew how they would react under pressure. So I believe that there's a certain aspect of mind over matter.

Of course, your choice of equipment reflects your personality as well, and I've done some fickle things. For example, I won two tournaments back to back in the 1980s with a putter that I found at a miniature golf course, with a lime-green tennis racket grip on it. If equipment feels good and looks good and you believe in it, then I believe you should use it. It's a matter of trying to find the equipment you can best dance with, so to speak.

The Golf Ball: Multiple Personalities

There's no more fitting symbol for the game of golf than a circle. Golfers complete their "round" on the 18th green just steps away from where they began it on the first tee. The ideal golf swing inscribes a circle through the air, and the hole and the golf ball are, of course, perfectly round. Far from a flat two-dimensional sphere, however, today's golf balls are full-bodied treasure chests of multiple technologies and performance tricks. Some, like Jack Nicklaus, feel the golf ball is *too* good now. They contend it flies too far and straight and is both making classic golf courses obsolete (for instance, because they have become too easy to play) and mid-level golf pros able to compete with the more talented players. Others feel that average golfers playing for fun deserve every advantage, aid, and inch afforded to them without violating the rules of the game.

Billy Mayfair, PGA Tour player, on today's golf ball

Everyone is talking about testing the new driver heads for illegal COR, etc., but I think the golf ball is making more of a difference in the game than anything else is. The golf ball has changed the game more than the driver has. I play the Titleist ProV1x ball, and it doesn't curve as much anymore like they used to. You just swing as hard as you want at it and it goes straight at the flag. Working the ball is still important on certain shots, but it's not as much of a must thing anymore. One of the guys I remember who used to work the ball the best in the world was the late Payne Stewart. He brought the ball in high, low, left-to-right, right-to-left, every which way possible. That was what made him such a great player. These days I don't think working the ball would help him as much, because you don't need to shape the ball like you used to because the ball just flies so straight. I think 95% of the guys on Tour would say that the changes in the golf ball today have made a bigger difference in the game than anything else has.

Just like golf clubs, golf balls must fit the golfer, and as with clubs, a player's task is to match the ball that best complements his or her specific swing and style of play. Golf balls, although not inexpensive, are affordable enough to allow a degree of experimentation. As they do with clubs, players should try different brands and types of balls before settling on one. Finally, golfers must match their golf ball not only to their swings, but also to their golf clubs. While doing so requires a bit of study and effort, the dividends it pays in better scores and shots can transform a round of golf from "a good walk spoiled" (to quote Mark Twain) into a day in paradise. In other words, it's a whole new ball game when it comes to the golf ball today.

Almost 10 years ago, the PGA Tour tentatively discussed a rule change that would have required all the pros to play the exact same brand and style of ball. The sentiment among players was that this would be fine as long as the new standard was the brand and style that they already played. Back then that meant a soft balata-covered wound ball (with rubber bands), such as Titleist's Tour Balata, or Maxfli's HT. The game's best golfers at that time categorically ignored the hard-feeling, two-piece/solid-core "distance" balls, such as Top-Flites and Pinnacles, favored by many average golfers. Although offering unmatched distance off the tee, these balls spun very little around the greens. The pros may "drive for show," but they chip and putt for dough. The image of one playing with a ball difficult to control in the scoring zone is as incongruous, if not hilarious, as a tennis player taking the court in snow boots.

The Top-Flite Strata golf ball, introduced in the mid-1990's, however, did the unimaginable. It merged a high-spinning, soft-feeling Tour Balata type of ball, with the low-spinning, long-flying, and durable Pinnacle or distance ball into a whole new class of product. This remarkable *three-piece ball*, was, indeed, two balls in one: It was a long-flying/low-spinning distance ball off the tee, and a high-spinning control ball off the irons. Remarkably enough, Strata's ball design team accomplished this two–for-one feat rather simply. They just added a soft polyurethane cover on what was virtually a Top-Flite distance ball, and then added a thin middle or mantle layer that encased the ball's already large and solid rubber core.

The Top-Flite Strata golf ball

A few years later, balls such as Callaway's Rule 35, Titleist's ProVI, Maxfli's M3 Tour, Nike's TA2 (both the Long and Spin models), and others, including new balls from Strata, improved on Strata's original breakthrough, by improving the durability of their urethane outer covers and making them thinner and firmer for added distance. At the same time, advances in rubber systems allowed ball makers to design cores that were more energetic or "faster," for even more distance on shots hit with the longer clubs, while maintaining a nice soft feel. Here's the short course on how these long-flying/soft-feeling balls work and why they represent a sea change in the history of golf ball design.

Titleist's ProVI golf ball

A golfer swings the driver, fairway woods, and long irons, on a relatively level path into the ball—so the clubhead penetrates through the three-piece ball's soft outer cover layer and compresses its firmer mantle layer and solid energy-packed core. On such swings, the clubhead "sidesteps" or mitigates the ball's high-spin-producing cover, which results in the kind of high launch/low-spinning drives that optimize distance (as discussed in the next section).

Furthermore, the reduced spin on the ball also means it will hook and slice much less than its Balata forebears, so the golfer gains not only distance but also accuracy.

On short and mid-iron shots, a golfer strikes down on the ball with a more descending blow. This action pinches the three-piece ball's thin and soft outer cover for shots that spin a great deal. The clubhead's force is also strong enough to reach or engage the ball's mantle level, which contributes height and distance to the iron shots. The blow with an iron, however, does not reach the ball's core or center layer, because the club contacts the ball obliquely at an angle, rather than squarely with all of its energy and mass as does a driver. If it had, the core layer's energy and speed would reduce the spin of the ball and it would be next to impossible for even skilled golfers to control the distance of their shots. Chips and putts almost exclusively use this soft cover for the kind of spin control and feel golfers need for scoring shots on and around the greens.

If the progression from a two- to three-piece ball yielded such performance benefits, why wouldn't companies progress to a four-piece ball? Of course, this is exactly what they did, with balls such as the Ben Hogan Apex Tour, Titleist's ProV1x, the Nike One, and the Strata Tour Ace. All but Titleist's ball feature a second firm mantle layer that acts like a conduit during impact that transfers extra energy into the core for even more low-spin-derived distance. The ProV1x achieves extra firmness by adding a second core. These balls perform best, however, for golfers with exceptionally high clubhead speed (in excess of 100 mph with the driver), because it takes considerable force to penetrate the additional material added to the balls.

Nike's One golf ball

At the same time, ball makers were applying this new fast/soft/low-spinning core and high-spinning/soft-cover technology in much improved (and considerably less-expensive) two-piece balls. Products such as Titleist's Next, Maxfli's A3, and others offered comparable distance as their three-piece compatriots, but marginally less spin off the irons and around the greens. This made them less

appealing to most Tour players and low-handicap-pers, although extremely popular with budget-minded mid- and high-handicap players.

Companies then softened the covers and cores of these two-piece balls and birthed yet another class of two-piece balls aimed at a different demographic. Balls such as the Precept Lady Diamond, Maxfli's Noodle Spin, and Nike's Power Distance Super Soft have given golfers with very slow swing speeds (some women, seniors, and juniors) more distance than ever. All of these two-, three- and four-piece balls are virtually indestructible. You can't cut them. You can't scrape them, you can't blemish them in any way. All you can do is lose them.

Although the superior multilayer balls have driven the pro and better amateur batty with delight (if not delusions of grandeur), they seem to have also intimidated average golfers, who feel they lack the skill needed to play with them. This is a tragic misconception according to Maxfli's senior director of research and development, Dean Snell, who, while working for Titleist, was instrumental in the original ProVl's creation. In fact, Snell believes the average player benefits as much if not *more* from today's multilayer ball than a pro. Here's why.

Dean Snell, Maxfli's Senior Director of Research and Development, on common misconceptions about today's multilayer golf balls

There's a big push in the golf equipment industry right now to develop new technologies. So custom fitting has become more important, and everywhere you go you see fitting carts and fitting centers. If a person is going to spend a thousand or two thousand dollars on clubs, they want to know about what they are playing, and they want to play clubs that fit them. A dozen golf balls costs 40 or 50 bucks, so people don't take the time to understand the technology or the difference between balls, and, consequently, don't usually play with the ball that fits them best. People today still think that if it's cold outside, they have to play a 90-compression ball, and if it's hot a 100, and that's a total myth. Compression or the relative hardness or softness of the old wound balata-covered golf balls no longer matters. Today we can make a ball with a large rubber center that feels soft like the low-compression balls of years past, but flies far with a lot of initial ball speed like the old high-compression balls.

But the real misconception average golfers have is that they feel they are not good enough golfers to play the new multilayer balls. They think that if Freddy Couples plays a Maxfli M3 Tour, for example, then they must not be good enough to play it, or that such a ball has too much technology that won't help them anyway. What the recreational player needs to know is that this technology is actually better for them than for the Tour player. It

helps the Tour player, sure, but it helps them more. Here's why.

First, these balls have such a low driver spin rate that they don't hook or slice very much at all, making it easier to hit the ball straighter with the driver. In fact, today's multilayer better balls have essentially the same spin rate off the driver as do the Pinnacles, Top-Flites, and other distance balls that they have been playing anyway. In other words, from the tee, these balls perform like distance balls, so there is no need to fear them.

Finally, when a Tour player shoots 70 in a round, he or she hits the driver 14 times, which we've fixed or made better because the ball goes farther and straighter for these 14 shots. Recreational players, who shoot 100, also hit 14 drives in the round, so they gain the same benefit off the tee as the Tour player does from the multilayer ball, with respect to this lower spin rate. Instead of playing 56 additional shots to the green and including putts (to make up their round of 70) as do the Tour players, they play 86 shots, which will fly higher, stop quicker on the greens, and offer more short game control and feel softer with the putter. That's 86 out of 100 shots that this type of ball improves for the average player, whereas for the Tour player it improves 14 out of 70 shots. So the percentage of improvement is actually higher for the recreational golfer than for the pro, which, again, means there is no need for average players to fear multilayer technology.

The Club/Ball Fitting Matrix: We've Made Contact!

The collision between the golf ball and the golf club, otherwise known as "impact," takes place in a fraction of a second. During that interval, the golf club transfers all of its kinetic energy into the ball, while its alignment at impact acts as "data" that programs the shot the golfer hits then watches with either self-admiring awe or disgust. The previous clubfitting section discussed the importance of fitting a player's golf clubs to his or her strength and swing style. To complete the clubfitting mission, and to optimize their performance on the course, however, golfers must also match their clubs to their golf ball.

Golfers have traditionally picked their golf ball by blasting drivers off the tee and picking the ball they hit the farthest. They generally paid little or no attention to the interaction between driver and ball or how that ball would perform with their irons, wedges, and putters. Today's new thinking about equipment fitting involves finding the best ball/club combination and posits that golfers should begin the fitting process for the golf ball *from the green back, not from the tee forward*. After they find the ball that performs best for them on and around the greens, they can then fit themselves for a driver that maximizes distance and accuracy with that ball as well. The following sidebar shows you how to do it.

Finding the Optimum Club/Ball Combination

Step 1   Go to the fringe around the green with a few (or several) new golf balls that you want to test. Hit some chip and pitch shots from different lengths and observe the results. The multilayer balls, such as Callaway's HX and Maxfli's M3, will come off the clubface at a relatively low angle. They will hit the green with considerable backspin and "check" noticeably before releasing and rolling toward the cup. Then try a two-piece ball, such as Titleist's Next, and notice how the chips and pitches fly a bit higher, check less when hitting the green, and roll a little more toward the club. (The super-soft low-compression balls, such as the Precept Lady Diamond, will roll the farthest, with the least spin when hitting the green.) Hit some putts and sand shots with these balls as well and observe their performance. Remember not every manufacturer's models of the same type of ball will react exactly alike.

Step 2   Take your same covey of balls and move out in the fairway to the 100-yard marker. Test each product from that point, and observe the trajectory, and the checking and releasing characteristics of each ball after it hits the green. Again, the multilayered balls will feel softer, fly a bit low, and stop or check more on the green than their two-piece counterparts.

Step 3   Now hit your test from the 150-yard marker and use the same criteria to evaluate each ball. The multilayer balls will spin the most and fly the lowest of the three, while biting more and rolling less on the greens. The two-piece balls will fly a bit higher and farther but spin and bite less on landing. Base your choice of ball on the combination of performance qualities that mean the most to you. A golfer who wants a soft feel, but also needs a little more distance, may decide to sacrifice the feel of the multilayer ball and choose a two-piece product.

Golfers who base their iron play on shots that hit and bite close to where they land would pick a multilayered ball, even if they have to sacrifice a little trajectory and distance. When you find the ball you like best, you're ready to find the right driver for it in the following way.

If you are lucky to find a launch monitor, look for the following launch conditions when testing a driver. Be sure to test with the same type of ball with which you plan on playing.

- A spin rate of between 2,500 and 3,000 rpms

- A launch angle of between 11 and 13 degrees

If your drives do not deliver these numbers, you can choose from a couple of remedies available. First, try a driver with a different loft: More loft will increase the launch angle and spin rate, and less loft will lower them toward this ideal range. You also can try a driver whose shaft has a stiffer tip end, which will lower the launch angle and spin rate, or one with a softer-tipped shaft, which will increase the launch angle and spin rate.

If you can't find a launch monitor, you can still fit a driver to your preferred ball by just your eyes. Look for drives that reach their apex or highest point very quickly and then level out and carry far down range. What you do not want to see are drives that start low and then shoot up like a jet plane taking off. Such shots indicate that the driver has added too much spin to the ball, which will result in shorter drives that will hook or slice more as well. Don't worry that your new high-flying drives will lose distance into a headwind. The wind doesn't blow any harder at 75 feet in the air than it does at 20 feet, and a low-spinning drive produces less friction against the wind than a high-spinning ball.

Jim Colbert, Champions Tour player, on the impact of technology on the Champions Tour

Jim Colbert   Photo Copyright Paul Lester

I don't think technology has had an impact on the Champions Tour as far as who wins at Tournaments is concerned, because everybody's got it. They're going to pass the money out at the end of the Tournament whether we shoot 20 under or 20 over. Having said that, the biggest difference I see in technology today is the ball.

Certainly it goes farther...My drives today are a little longer than when I was on regular Tour...but mainly it's the ball that doesn't curve as much as the older balls. Subconsciously, though, the curve of my shots is still in there, because at times I find it difficult to aim dead straight. I've gotten so accustomed to seeing the ball move left-to-right or right-to-left, depending on the shot I want to hit. And the balls today *really* don't curve when you hit them solid.

I don't think it's a good idea to have two separate golf balls, or two equipment standards, one for the Tour pros and another for the amateurs. Commercially, that would be a big mistake and would take a lot of money out of our industry. Not only would the pros lose their endorsement contracts with the equipment makers, but also the average player would lose the benefit of our testing the products they eventually buy and play with. The Tour really is the test ground or laboratory where the manufacturers develop their new clubs and golf balls. Now if they want to slow down the ball a little bit so it doesn't go quite so far, I suppose that would be okay as long as everyone plays with the same slowed-down ball, pros and amateurs. But you have to remember that Tour pros are a very, very small percentage of the golfing population, so people are making a big deal over a small group. Anyway, I like to see our customers, meaning amateur players, get every advantage technology can give them.

Golf Shoes: From the Ground Up

The first bit of wisdom many of golf's best teachers share with their students is that a good golf swing starts from the ground up. Indeed, today's golf shoe makers heed this call with a host of technologies that work bio-mechanically with a golfer's swing to augment traction and balance for more solidly struck shots. Just as a golf club integrates the grip, the shaft, and the clubhead into a smoothly swinging unity, a golf shoe's triumvirate of traction, comfort, and fashion help golfers play, feel, and look their best.

It wasn't long ago that golfers resisted the change from metal to soft spikes because they feared a loss of traction. When they began to experience the benefits of putting on greens not torn to shreds by metal spikes, they turned to the softies in droves. Today, virtually all golf shoes come only with soft spikes, which almost every golf course in America requires its players to wear. What's more, improved injection-molded thermoplastics have made these spikes more durable than in years past, just as research and development has improved their traction.

Billy Mayfair, PGA Tour player, on today's golf shoes

Golf shoes have gotten better in terms of their support, which helps Tour players play more than 18 holes in a day if they have to or to play more days in a row. They certainly have made my feet feel better; for instance, I don't get blisters on my feet as much as before. I think the biggest issue with golf shoes is the change from metal spikes to the soft spikes, and not so much because soft spikes don't tear up the greens, but because they are so much lighter to wear. They have helped my legs and back tremendously just because they are so much lighter, and part of that is their technology, but a big part is the soft spikes. When some-one sends me a pair of golf shoes with metal spikes in them now, and I go out and play one round with them on, because I'm used to soft spikes, my feet feel like they have bricks on them.

The top golf shoe manufacturers now use newly developed twist-and-click soft spike installation and extraction technologies, such as Q-LOK and FastTwist. This technology works with a simple and swift twist of the wrist and makes the old way of twisting metal spikes on and off with a horribly

Footjoy's Classics Dry Premiere golf shoe

Nike Golf's Gore-Tex TW Cap Toe golf shoe

designed wrench seem as antiquated as the Model T. With names such as the ChampScorpion and the Softspike's Black Widow, these spikes are softies yet nevertheless deliver serious ground-gripping bite. What's more, their well-designed wide bases also mean they clog with dirt and debris less than their earlier versions.

At the same time, golf shoe makers such as Nike, Footjoy, adidas, and others began designing durable high-tech polyurethane outsoles on their products. Fitted with built-in ridges, nubs, wedges, and edges and designed to add traction by working with the removable soft spikes, many of these outsoles offer more than 150 points of contact with the ground. All grip and release the turf in a predetermined sequence and at the proper moments during the golfer's swing.

Callaway Golf entered the shoe market with its patented positional traction technology, which includes the oversized, asymmetrical, and large Big Bertha soft spikes. These spikes have hard and soft halves oriented in specific directions and a twist-and-click installation method that enables golfers to replace the spikes in their shoes in the same orientation every time.

Callaway Golf's Big Bertha spikes

Companies such as Footjoy understand that a golf shoe must do more than more provide stability during the swing: It must also roll and flex to accommodate the weight shift and motion of the body's pivot. Indeed, Footjoy's GelFusion shoes, with their innovative iSuspension outsoles, have toe and heel portions that flex independently for optimal contact with the ground regardless of the terrain's slope. adidas' premium Z-Traxion series shoes accomplish a similar kind of dynamic flexibility by using their torsion bar, which runs down the center of the outsole, to allow toe and heel to work separately.

The adidas Z-Traxion Tour

Manufacturers know that their technology goes to naught if their shoes don't feel and fit well, and they have gone to many high-tech measures to ensure that they do. For example, the adidas Tour Performance series features a FitFoam lining that molds itself to the conTours of each golfer's foot. Nike's entire line of golf shoes has lightweight gas-inflated cushion sections positioned in the shoes' heels and (in some models, the fronts) for a velvety-soft walk.

Park City, Utah-based Surefoot, a high-end ski boot retailer, has gone digital with their custom-made golf shoes. Their fitting process begins with a 360-degree computer-scanned image of the golfer's feet, with each foot measured in 538 places. From this data, the company builds each person's welted leather pair of shoes one at a time, which include milled orthotic inserts for added support and stability.

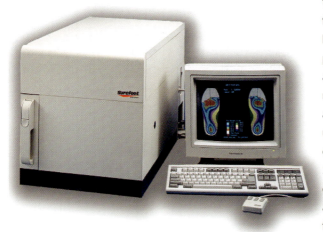

The Surefoot computer scanning system

Shoemakers have also made significant strides in keeping the feet inside of their products dry. Whereas in years past waterproof liners were heavy, sweaty, and hot, today's advanced Gore-Tex linings (and other similar fabrics) keep moisture out and feet cool and airy. Many high-end golf shoes add moisture-wicking fabric liners that move perspiration away from the feet.

Speaking of air, a couple of years ago a company called Bite introduced golf sandals, with soft spikes installed on the soles. Although ridiculed at first, golfers with arthritic or achy feet have taken to them unselfconsciously, especially since Footjoy legitimized the concept with golf sandals of its own. The new adidas ClimaCool Slingbacks for women blend a sandal-strapped heel with a meshed upper front in a shoe that weighs 9 ounces.

FootJoy's golf sandals

Having raised golf shoe technology to a higher level, shoemakers have given their designers a free reign to explore a new range of styles. Golfers can choose the solid formalism of welted Footjoy Classics, with fine calfskin leather uppers and water-resistant leather/thermoplastic outsoles. They may want to push the envelope with a sense of youthful exuberance with shoes such as Oakley's Bow Tye, whose molded synthetic rubber outsoles kind of hip-hop up and blend into the saddle portions of the leather uppers. Women can walk boldly into the same lascivious red Veranda Slip-On's from Nike that Annika Sorenstam once wore on Tour, while the khaki crowd might go for one of the many lightweight and laid-back tennis-shoe-style golf shoes available from most top manufacturers.

Doug Robinson, FootJoy's Senior Director of product development and advanced concepts worldwide, on the Techo-Packed GelFusion shoes from FootJoy:

The GelFusions have a tremendous amount of technology built in to them, which I'd like to outline and explain for you.

The iSuspension outsole on GelFusion presents a dynamic technology where the forefoot and rear parts of the shoes rotate independently. You can see on the shoe's bottom that the two halves of the outsole are joined together with a TPU (thermoplastic urethane) ball-and-socket joint that snaps together and allows the foot to work naturally, very much the way it does without a shoe. For better mobility, we've carved out the midsole of the shoes to, again, help both halves of the foot rotate more freely.

The GelFusion also incorporates OptiFlex technology for optimum forefoot flexibility. A softer TPU material is molded into the outsole in the area of the metatarsal heads, or the "knuckles" underneath the forefoot. This allows this very dynamic and active section of the foot to freely move while walking or swinging a golf club.

One of the many gel technologies in the GelFusion shoes is found in our GelRide feature. We developed a lightweight composite gel that is molded into the heel of the midsole for additional walking comfort.

The GelFusions present a very advanced nonmembrane waterproofing system that took four years to develop with our major leather vendor, Pittards of England. We call it the AQUAf.l.e.x system. In essence, we've tanned and fused, at the fiber level, weatherproofing chemical agents directly into the leather upper and linings of the shoes themselves. This provides the waterproofing capability of a membrane, but with more breathability and suppleness and without a membrane's weight.

There is also in-shoe climate control technology that employs a material called Dryz IntelliTemp. This high-tech polymer has two unique technologies incorporated into a comfortable cushioning foam, moisture management and cooling. First, the Super Absorbent Polymer (SAP) in the foam pulls moisture away from the foot and gels it to keep the foot dry. When golfers take off their shoes, the moisture evaporates and this material returns to its original state, ready for the next round. Second, the Dryz IntelliTemp uses microsphere Phase Change Materials (PCMs) that act like a thermostat, pulling heat away from the foot when hot and returning stored heat when cold. We use Dryz IntelliTemp in the footbed and tongue. The collar of the shoe is a molded gel combined with the PCM, in what we call the GelCollar. Normally, there is just foam in this area. The GelCollar, molded to the shape of the foot, provides excellent fit, comfort, and additional cooling.

We want golfers to arrive at the 18th tee with feet that aren't tired and legs that are fresh. Our

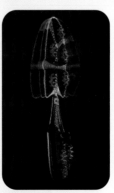

goal with these and all of our shoes is to provide very stable hitting platforms and the best blend of traction, waterproofing, and comfort. That's the bottom line in terms of golf shoes, and that's what golf shoe manufacturers will keep trying to improve on in the future.

The GelFusion technology

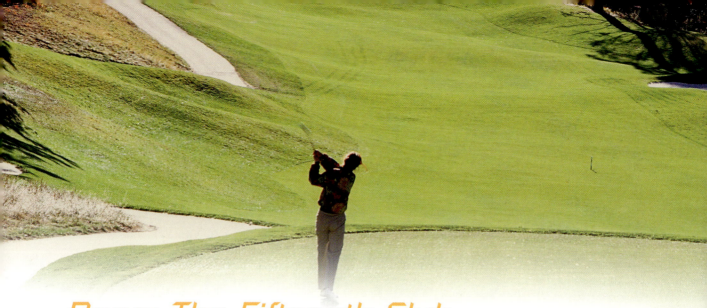

Bags: The Fifteenth Club

If golf is a journey, then those who play had better pack their bags well. That's easier to do today than ever, with a greater array of new feathery-light bags blazing with candy-store colors, copious compartments (for jewelry, cell phones, water bottles, peanut butter and jelly sandwiches?), and more technological treats than ever. In fact, because golf bags present none of the frustration of hitting bad golf shots, everyone can readily partake in a tension-free enjoyment of their technology, features, and attractive designs. Yet, for those walking the course, lightweight bags qualify as pieces of golf equipment, because in reducing golfer fatigue, they can contribute to lower scores.

Today's crop of bags fall into two basic categories: cart and stand bags. Designed for the back racks of motorized golf carts, the cart bags are larger and heavier than their stand bag counterparts. This discussion covers stand bags exclusively, which generally offer the same technology and features as cart bags.

Whereas opening a stand bag once felt like a wrestling match with an angry octopus, today's versions work with the smooth grace of Fred Astaire dance step. For example, Sun Mountain's lightweight RollerTop Lightning bag lets golfers grab a plastic handle attached to the rim, swing the bag off their shoulder, and extend its legs onto the ground in one fluid motion. Sun Mountain's Superlight 2.5 may also offer the most feathery hike available, weighing in at a svelte 2.5 pounds. When a stand bag finds the ground, innovations such as PING's hoof-shaped base on its Hoofer 3 stand bag take over with a solid base to keep them from toppling over.

Golf bags from PING and Sun Mountain

The ever-innovative OGIO (see company CEO Michael Pratt's sidebar) offers the Exo bag (standing for exoskeleton), which features the Woode club management system that strategically separates woods and irons into full-length dividers for optimum carrying convenience and balance. Organization buffs will appreciate Club Glove's (from West Coast Trends, Inc.) Collegiate II's fully modular side pockets, which easily snap on or off, depending on what the golfer needs to pack for a round.

Mizuno's AeroFrame

The Collegiate II

In years past, a lightweight bag meant a bag about to fall apart, but that has changed. Mizuno's AeroFrame, for example, features an aluminum frame construction that can bear a golfer's entire body weight without bending or breaking.

Virtually every carry bag today comes with a dual shoulder-strap system either designed by Izzo, the

company that invented it, or one based on it. These types of straps distribute and balance the bag's weight comfortably across the golfers' shoulders and back. Nike augments comfort by bolstering the Izzo straps on their line of stand bags with an Air Cushioning system similar to that used in their golf shoes.

The Club Glove Last Bag from West Coast Trends, Inc., is an extremely high-quality and durable travel bag used by a large number of Tour pros. Made from Dupont-developed nylon and featuring inline skate wheels with bearings, the bag rolls as easily as the golf traveler walks to the airline ticket and car rental counters. The company also makes the Kaddy Stroller, an easy-to-fold push or walking cart with a familiar three-wheel design based on baby strollers (see the sidebar). Golfers who don't want to pull or push anything but their competitors' envy buttons, might opt for the Hill Billy, or Lectronic Kaddy, two of several fully motorized (and expensive) push carts on the market.

Michael Pratt, CEO of OGIO, on the future of golf bags

Mike Pratt, CEO of OGIO, and the OGIO Exo bag

I think you're going to see a lot more technology in golf bags from companies like OGIO and others that focus exclusively on making bags. There haven't been a lot of innovations in the technology in golf bags. We've improved on materials and made better functioning kickstands and so forth, but not much that's really new, but it's coming. Take our new Hoode system here at OGIO. It's a retractable spring-loader rain hood that is built in to the tops to the bags themselves, and it pops over to protect the clubs from the rain.

In the future we'll see the shift continue from clunky heavier bags to ones with improved lightness and sturdiness. We'll do this with better plastics and tooling innovations that will make the component parts very light and also very strong. I think we'll see more innovation in carrying systems that will make walking much easier, because a lot of people are frustrated carrying a golf club around for 18 holes. The goal from manufacturers will be to make more comfortable carrying devices that are friendly and simpler to use. When we do that, more people will walk, because they want to walk.

Cart bags will become easier to put on the golf carts, and it will be easier to get into the pockets on those bags. We'll also start to see more stand or carry bags that double as cart bags. Right now, it's difficult to fit the retractable legs of a carry bag onto a cart. The products are starting to come.

The Club Glove Kaddy Stroller from West Coast Trends, Inc.

The Kaddy Stroller

Although designed with a baby stroller in mind, this cart's litany of technology would probably appeal more to Lance Armstrong. It includes the following:

- An adjustable, padded handlebar and a lightweight aircraft powder-coated alloy frame and high-impact plastic components for single-action fold
- Front and rear quick-release pneumatic treadless tires with metal valve-stem extensions

- Rustproof alloy rims and sealed steel ball bearings
- Front wheel handbrake that provides speed control down hills
- Parking brake along with foot-pedal-activated rear wheel brakes
- Adjustable nylon and rubber security belts that provide a snug fit for nearly any golf bag
- A multifunction organizer, water bottle holder, and a tire pump for the cart's inflatable tires

Hopefully, the golfers' games will be as efficient as this cart they are pushing.

Gloves: A Helping Hand

Golf is about making connections. The feet connect to the ground for traction and leverage during the swing. Friends, family members, and business associates connect with one another during a round. The golfer connects with nature and the great outdoors, and the hands connect with the golf club through the grip to swing the club with control. For some serious bonding with their clubs, golfers often turn to golf gloves.

The classic Cabretta leather glove comes from the thin hide of sheep found in very arid countries such as Ethiopia, Nigeria, and Sudan. It offers a soft yet tacky hold on the club, and remains the choice of many, if not most, Tour professionals. Historically, these gloves offer little in terms of extra or "visible" technology: They are just well-made products that fit the hand snugly and offer the rich, unmistakable aroma of fine leather. Their liability has always been their relative lack of durability and breathability and their high price. Glove manufacturers, however, have improved these gloves with an array of technologies to extend their life span, bolster their flexibility and fit, and lighten and air them out.

Gloves made from synthetic nylons suggest the look of leather and a lower price point, while still providing good performance. For a higher-performing compromise, golfers can choose a hybrid glove, which combines Cabretta palms (or Cabretta palm and thumb patches) and synthetic back portions for durability, coolness, and better fit. In fact, several Tour pros now use such gloves, mostly because of their improved fit.

Here is a look at some of the best golf gloves available today.

FootJoy's SciFlex gloves

Tour Cabrettas

FootJoy's SciFlex snappy high-tech look might attract Spiderman wannabe's, or at least golfers looking to secure their hold on the club. It juices up the traditional Cabretta leather by impregnating it with microfibers that either solidify or liquefy, depending on the temperatures of the hand and outside air, for maximum comfort. Cooolmax Lycra across the knuckles and a PowerNet mesh along the three-dimensional ComforTab Closure combine to enhance the glove's flexibility and fit. The same company's SpiderGlove, used in very humid weather, integrates graphite into the Cabretta for better gripping, durability, and moisture control.

Mizuno's Soft-Fit is a techy beauty made of fine Cabretta leather. Its BIO-LOCK fitting design works to ensure a new glove fit each time the golfer puts it on with its innovative Y-shaped closure system. Lycra panels in key areas improve feel, fit, and flexibility, and a double-stitched thumb prevents twisting and bunching for the life of the glove.

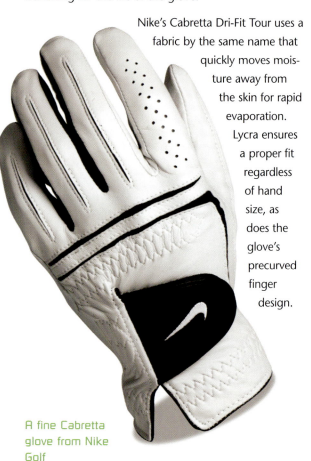

Nike's Cabretta Dri-Fit Tour uses a fabric by the same name that quickly moves moisture away from the skin for rapid evaporation. Lycra ensures a proper fit regardless of hand size, as does the glove's precurved finger design.

A fine Cabretta glove from Nike Golf

Titleist's Player's Glove has become one of the Tour's favorite gloves. It's a no-frill product that uses extremely thin Cabretta leather for the best feel, quality design, and precisely placed sewn seams to produce a consistent fit and flexibility.

Synthetic and Cabretta Hybrids

Etonic's synthetic AC FEEL and other gloves feature Cabretta palms and microfiber synthetic, nylon-based back sides that apply air-cooling technologies for moisture and heat control and for a better fit.

This glove and others, such as Footjoy's, add additional Cabretta leather patches in the thumb and palm areas for extra gripping and durability in these high-friction zones.

Footjoy's RainGlove offers a very good pair of non-woven microfiber gloves that expand like a sponge when wet and conform to the golfer's hand. This glove grips so well when saturated with water that the manufacturer actually suggests wetting it before use. Golfers looking to extend their rounds beyond the autumnal equinox can buy a pair of winter gloves reinforced with lightweight fleece

TechTip

Maria Bonzagni, Senior Director of Marketing, Footjoy Gloves and Accessories, on three simple tips for caring for golf gloves

1. As with golf shoes, it is best not to wear the same golf glove two days in a row.

   Alternating between two or more reduces wear and tear and greatly prolongs the life of your gloves.

2. Make sure you keep your glove flat between your rounds, so that it maintains its form and shape.

3. Keep the glove inside the envelope it came in from the manufacturer or place it in an airtight plastic bag. You can keep the glove in your golf bag, as long is it lies flat and fully formed in its envelope or bag.

Golf Gadgets: For Golfers Who Thought They Had Everything

The dictionary defines the word *gadget* first as "any small contrivance," and then as "some little thing whose exact name is unknown or forgotten; thingamajig." Golf gadgets, indeed, come in many odd shapes and forms. Unlike a swing or short-game-training aide, however, they do not work directly on your technique, but offer a creative helping hand in unexpected and imaginative quirky ways, to make golf less awkward, more fun, and, most importantly, easier to play. Of course, implicit in their quirky designs and conception is the idea that by using them, golfers' scores will improve as well. At their hearts, whether high tech, low tech or no tech, they are primitive tools, like the twig the chimpanzees stick into the termite mound for a savory snack. As such, they are not perfect: In fact, most golf gadgets have an annoying flaw that discourages their widespread use. However, here of some good ones just the same.

The Golf Ball Pick-Up Golf Ball Retriever

This may be the prototypical golf gadget. It is just a small suction cup that slips over the top of the shaft of any standard putter. Instead of bending down to get the ball out of the hole, you can use this grabby friend to pick it right up, thus saving expensive trips to the chiropractor's office or physical therapist's clinic. The problem with it is twofold. First, it adds a bit of weight to the grip end of the putter, which makes the clubhead feel a bit lighter. This may disrupt the timing of the stroke and it can also rub annoyingly against the golfers' forearms or torso, depending on their putting style.

The Technasonic Check-Go Sweet Spot Finder and the Line M Up Golf Ball Alignment System

The theory behind the Technasonic Check-Go Sweet Spot Finder, which sells for about $25, is that no golf ball is completely round. Golfers place the ball on top of a small metal platform, push a button, and watch the ball rotate like mad. Supposedly, the high-speed gyroscopic action realigns the ball's variance in mass along the equator, which the golfer marks with a small felt-tipped pen while the ball spins. Golfers should aim this line straight at the hole while putting and along the target line while driving, which ensures that the ball will roll and fly more steadily and true. Unfortun-ately, one hardly ever putts from a perfectly flat surface, which mitigates the alleged benefit of the centered ball. As for driving, even the game's best say that the straight shot is the hardest of all to hit, so the driver almost always imparts some sidespin on the ball regardless of its orientation on the tee. But it's a fun and dizzying spin on the ideals of the straight and narrow just the same.

Golfers with vertigo can avoid the spinning with another simpler alignment gadget called Line M Up. After placing the ball in a plastic encasement, golfers trace a line around the ball through guide slots. This one doesn't indicate the ball's balance axis, but golfers can use the line as a good aiming aid while putting.

The Divix Divot Tool, Putter Holder, Ball Marker, Cigar Punch

Golf is a game of etiquette, and one of the first cour-tesies new golfers have to learn is how to repair their divot marks on the greens. So important is this activ-ity that many golf courses give each player a metal two-pronged divot tool for free. This one costs closer to $15, but, hey, look what a golfer gets! First of all, it springs open and retracts with the push of a but-ton like a switchblade. (Take it out of your pocket when going through metal detectors. Actually avoid carrying in your pocket at any time; these things have a nasty habit of stabbing golfers in their thighs when carried in a pants pocket.) When chipping

from off the green, golfers can stick the Divix into the ground and keep their putter dry from wet or dewy grass by resting it on the cradled end. Removing the nickel-plated ball marker reveals a cigar punch. The only problem with this or any other divot fixer is getting golfers to use it.

Shot 2 Golf Ball Cleaner (from Shot 2 Incorporated)

One of the jobs a good caddy performs is cleaning their golfers' balls carefully and thoroughly on the greens, so dirt and debris don't interfere with the roll of the putt. When playing without caddies, golfers can clip the Shot 2 Golf Ball Cleaner to their hip pockets and, with one hand, wash their golf balls on a compressed sponge and dry it with a pocketed towel. The gadget rests in a comfortable and unob-trusive position that doesn't interfere with the golf swing. The problem is that the attaching device can be a little cumbersome to use

The Brush-T (from Brush-T Innovations, Inc.)

When a driver contacts a golf ball perched on a tee, the ball both flattens and elongates. As it reconsti-tutes into its original shape, it releases a great deal of energy that contributes to the distance of the shot. As it compresses, it also pushes down against the top of the wooden tee. This creates a bit of friction, which adds spin to the ball, and excess spin reduces distance and contributes to hooks and slices. The Brush-T tries to remedy all of this. Golfers place their ball on a group of firm bristles, which offer very little resistance against the ball at impact. Less resistance means less spin for potentially more distance and less hooks and slices. The tee comes in three heights to accommodate tee-ing when playing irons, three woods, and today's deep-faced drivers. Still, some golfers with

The Divix Divot Tool

The Brush-T at work

slower swing speeds need as much spin as they can get to keep the ball flying high in the air, making this gadget better for strong golfers with fast swings.

Groove Cleaner

This gadget approximates the simplicity of the chimpanzee's termite twig, and, at less than $2, it is almost as cheap. Its metal tip at the end of a hardwood handle effectively cleans the grooves of golf clubs to maximize their performance. Traditionally, golfers use the tips of wooden tees to get this job done, but these tend to wear down quickly. Instead of efficiently unclogging a club's grooves, which ensures shots fly and "bite" the green with sufficient backspin, a blunt tee trying to scrub clean an iron's or wedge's grooves does little more than raise the ire of a frustration-saturated golfer. Maybe the Monkees should rerecord Simon and Garfunkle's "59th Street Bridge Song," whose refrain is "feelin' groovy."

Body Kool, Reusable Bandanna (distributed by Rose Industries)

Golfers playing in very hot weather will appreciate this bandanna, which features high-tech polymer beads sewn into the material itself. They absorb water and expand to hundreds of times their actual size, and as the water evaporates from the crystals the bandanna cools the body. Golfers activate the bandanna by soaking it in cool water for 25 minutes and/or placing it in the refrigerator for extra comfort on hotter days.

Golf Gadgets from Brookstone

The king of gadget retailers offers a couple of good golf ones, including a clip-on golf-themed quartz Golf Watch, which attaches with a strong metal clip to a golf bag. Golf nuts putting in long days on the job can slip out for a few twilight holes (perhaps segueing into night) with the Twilight Tracer Light-Up Golf Ball. Two red LEDs flash 7.2 times per second to help you locate your ball, which, although regulation as far as distance limits is concerned, is off limits as far as the Rules of Golf are concerned. When playing by the rules in competition, on the other hand, golfers must mark their golf ball to identify it as theirs, and they can do so with the Golf Ball Personalizer. This stainless steel rubber-handled gadget has an easy-to-use 26-lettered monogramming tool that can stamp up to 3 initials clearly on the ball. The kit comes with ink and lacquer.

Greg Melton, of TechTV, on golf gadgets

Golfers look for anything to help them shave just a few strokes off their game. How many late-night infomercials have you seen pitching some revolutionary product that will turn an ordinary golf swing into one a pro would be envious of? In honor of Wired World of Sports Week at TechTV, we compiled a list of golf gadgets that will help you lose a stroke or two off your game.

- Putt 'n' Hazard with Breaks—Complete with one sand trap and water hazard, this device will help you hone your putting skills without having to drive to the course. You'll also be able to simulate and set right and left breaks via remote control. Don't worry, it even comes with an automatic ball return.

- Electronic Driving Range—This computerized driving range measures the length of your shots and provides instant feedback on distance and accuracy. The best part is that the ball is always ready for the next swing.

- Golf Scorekeeper/Pedometer—That's right, you can keep score of how many strokes you've swung as well as how many strides you've taken throughout the day. It also features a clock to keep you on pace.

- Slip on Spikes—Slip on spikes may or may not help you improve your golf swing, but they are a great thing to have in your bag if you've accidentally left your golf shoes at home. These shoes attach directly to a regular pair of sneakers.

- Gold Score Card—This little digital scorekeeper tallies up to four players' scores over 18 holes and displays them on the leader board.

- Schmeckenbecker Putter—This is the ultimate golf putter. It features a compass, a candle for the 18th hole, a rabbit's foot for good luck, and an air horn to alert others when it's your turn to putt.

The Golf Course
Perfecting Paradise

The legendary teacher Harvey Penick instructed his students, including greats such as Ben Crenshaw, Tom Kite, Mickey Wright, and Kathy Whitworth, to "Take dead aim," and in so doing captured the essence of golf as a target game. Today's vastly improved equipment certainly deserves all the attention it receives, as do the game's teaching gurus tips in the magazines. Golfers lose touch with the essence of the game, however, if they forget that its joy lies in negotiating their ball across that bucolic kingdom of obstacles otherwise known as a golf course.

The men and women who design, build, cultivate, and care for golf courses never forget this. The golf course architect strives to build the ideal course like a sculptor carving stone from a block of marble. The agronomists' search for the healthiest grasses mirrors the medical researchers' quest to cure disease, and the golf course superintendents care for their courses as if they were children. In short, a golf course stands as a hallowed if not sacred zone separated from the distractions of everyday life, where the gamut of human emotions play themselves out in a sequence of always surprising adventures known as a round.

Advances in technology and tools used to design, build, and maintain golf courses have done much to improve both the golf course and the game itself. At the same time the golf course today finds itself in the middle of several controversial issues. Can architects still design courses with the aesthetic and shot-making values of classics courses, such as Merion, Riviera, St. Andrews, and Pebble Beach? Can they retrofit such courses to accommodate the new long hitting clubs and balls? Has contemporary golf equipment reduced new courses to behemoth open parks, whose fairways extend with the endless length of an airport's runways, and whose closely cropped greens require the flat grading of a dull table top?

And more: Should the PGA Tour turn to a "tournament ball," which does not fly as far as today's golf balls, so that classic courses will become challenging again, or does the integrity of the game require all players to use the same class of equipment? Can developers build courses that not only do not negatively impact the environment, but also enhance it in some ways?

While technology may have created some of these problems, it is to technology that the golf course professionals turn for more effectual answers and more creative solutions.

Today's Golf Course—The Playing Ground Redefined: How Contemporary Equipment Has Changed the Shape of Golf Courses Old and New

A fundamental conflict frames the discussion of today's improved golf equipment and its impact on golf course architecture. On the one hand you have the manufacturers who want to design easy-to-hit clubs and balls that go forever to make the game easier for the average player. At the same time, golf course architects try to mitigate the advantage gained by using such gear by designing longer courses (or retrofitting old ones with more length) that challenge the game's best (that is, the Tour pros). Perhaps, ironically, the only thing that seems more insatiable than the average golfer's desire for more distance is the golf course architect's commitment to defending golf courses' honor and integrity from young flat-bellied and limber-backed pros whose long-distance games threaten them.

In their attempt to do this, golf course architects, such as John Fought and others (see sidebar), now both design new courses and retrofit old ones to what only a few years ago would have seemed like ridiculous yardage lengths.

Southern California-based golf course architect Steve Timm explains the need for these longer courses. He points out that although the new drivers hit the ball farther, they don't always hit it straighter. Timm says that the guidelines set by the Federal Urban Land Institute recommends that a home or building sit at a minimum of 210 feet from the center line of a fairway. However, strong golfers armed with a 400cc, or larger, driver and multilayer golf balls can just as easily hit a ball 260 yards off-line to the right or left as dead straight. The screech of a window shattering on a house buttressed up against a fairway has become an all too familiar sound to golfers writhing in shame and embarrassment on the tee. The only solution, Timm suggests, seems to build longer and wider golf courses.

Thankfully, as standard practice, architects such as Timm fit these courses with several sets of tee boxes, which allow golfers of varying strengths and skill levels to enjoy and compete together on the same course. Conventionally, the handicapping system balances the differences in skills between two players by giving the less-able golfer a certain number of strokes per round. Today, stronger players may have to offer distance *and* stokes to their weaker opponents to level out these fortified playing fields. Even so, some golf course architects, such as Robert Cupp, have begun experimenting with ways of building shorter golf courses that do not humiliate players with their degrees of difficulty yet preserve the game's challenges just the same (see sidebar).

Bigger golf courses raise other issues, both positive and negative, that go beyond those concerning professional competition. First, more land means courses that cost more to build and maintain, which inevitably translates into higher greens fees for golfers. Bigger courses take longer to play, and slow rounds

represent an ongoing problem in the golf industry. On a more positive note, larger courses also spread their trees farther apart, and this reduces shade, both letting the grass grow better and improving ventilation for the course and for golfers. What's more, some golf course architects welcome the opportunity to express their design ideas on a larger canvas, and many golfers have come to enjoy the new grandeur in feeling and look of today's bigger courses.

Today's improved lawn mowers have turned greens, and even fairways, into friction-defying freeways as slick as hockey rinks. As a result, golf course architects rarely, if ever, design greens with the kinds of severe and dramatic slopes as they did in years past. Instead, today's large greens display segregated sections with subtle slopes and undulations surrounded by flatter pin placement areas. These not only present fairer putting surfaces for golfers, they also give golf course superintendents the needed flexibility in selecting pin locations on greens that fully integrate the design and strategic features of the holes.

A golf course's grandeur as seen from overhead

John Fought, golf course architect, on technology and creativity in contemporary golf course architecture

The distance the Tour players now hit the golf ball has changed the game tremendously, and, at its worst, has made the wonderful old golf courses such as Merion and Cherry Hills completely obsolete for PGA-Tour level tournament play. I'm not worried about the Tour players and the low scores they shoot, because they are all such awesome golfers that they will break par on any course, long or short. The problem is that the way the game is played on Tour today, the course no longer forces the players to *think* their way around the layout. When they're playing a great course such as one of the ones I mentioned, I want to see them hitting the same kind of middle and long iron shots to the par 4s that the Hogans, Sneads, and Bobby Joneses did when they played these courses. The equipment today has made all the par 5s into par 4s, and it's made almost every par 4 into a drive and a wedge shot. We used to have real par 4s and I'd like to see them come back!

John Fought designed Windsong Farm, in Minneapolis, Minnesota

When I build courses for clients who want to attract a Tour event, I build them with an extra tee that lengthens the course to as much as 7500 yards, which, of course, is too long for almost all the course's members. Now if we started designing all the new courses at 8000 yards to challenge the pros, we would make the wonderful classic shorter courses such as Merion seem second rate for Tour tournament play, and that would be an awful shame. Not all the classic courses have the land available to make them longer, although some do, and it can, has been, and will continue to be done.

Look at Augusta National, for example, home each year of the Masters, where the lengthening of the course has brought back into play some of the hazards as they were intended for the great players of past eras. I remember watching Tiger Woods blasting his tee shot over the fairway bunker on the right of hole 8 at Augusta just this year. Before they lengthened that hole, he would simply carry his drive right over the fairway bunkers on the right. The ball would land in a part of the fairway that is twice as wide as it is in front of the bunker, which was the originally designed landing area. By moving the tees back, Tiger could no longer fly the ball over the bunker. Instead, he had to play his tee shot to the left of the right fairway bunker the way Dr. Alister MacKenzie wanted the hole to be played when he designed the course in the 1930s.

Look at another great classic course like Riviera, which also has been lengthened recently, but still isn't excessively long by contemporary standards. Every shot there holds the golfer's interest, every hole is both different and memorable, the types obstacles change throughout the course, and there is imaginative putting. So is length important? Yes. It's part of a contemporary golf course's equation. But it's not everything.

Geoff Shackelford, author of *Grounds for Golf: The History and Fundamentals of Golf Course Design* and the co-architect of Rustic Canyon Golf Course in Moorpark, California, on why he favors a "Tour" golf ball

There are obvious advantages to establishing a Tour ball that wouldn't fly as far as those that amateurs would use. The often cited one is that architects could stop lengthening the existing classic courses and the golf fan would get to see Tiger play the same courses on which Snead, Hogan, Trevino, Palmer, and a young Jack Nicklaus won. They also would see and enjoy much more shot making on the pro level, because the Tour-designated ball would spin more and spin on the ball is required for shot making and shot shaping. The reintroduction of shot making on the professional level will lead to a more exciting PGA Tour, which will breed interest and enthusiasm in golfers who will then stay with the sport, instead of dropping the game, as many do now. Many truly avid golf fans can't stand to watch a Tour event these days because the golf has become boring. A few years ago, before the introduction of some of today's outrageously long golf balls, these were the same people who taped the final rounds of tournaments every week.

The argument against the "Tour ball" is that average golfers not using one won't get to compare themselves to pros when they play a Tour tournament course. However, I predict that as soon as a Tour or competition ball came out on the market, most golfers would buy a version of it made by their favorite manufacturer just to try it out. Some golf clubs and local tournaments might make the ball mandatory, which would only increase its sales. So, clearly, I see this negative quickly becoming a positive. Not only would golf ball manufacturers have one more product to sell on the shelves, owners of shorter golf courses also won't have to waste money making themselves longer to attract Tour tournaments. That savings can be passed on to the golfer in the form of lower green fees, and less-expensive golf can only help the future stability of the sport.

Wally Uihlein, Chairman and CEO, Acushnet Company (makers of Titleist and Pinnacle Golf Balls), on why he believes the game has not been harmed by advancements in technology

Wally Uihlein

The 85th PGA Championship recently contested at Oak Hill will be remembered as one of the greatest finishes in major championship history. Any golf fan watching will never forget the 7 iron that Shaun Micheel hit to 2 inches on the 72nd hole to secure not only his first major victory, but the first win of his PGA Tour career. Ironically, it was at the same site in 1968, 35 years earlier, when a 30-year old Texan, Lee Trevino, accomplished a similar feat, earning his first PGA Tour victory and first major champion title, the U.S. Open. His performance included becoming the first player to shoot four sub-70 rounds in U.S. Open history, and his 275 total was one shot better than Micheel's winning score this year.

With the four majors complete and two months remaining before the Tour Championship, an interesting thing has happened to professional golf in 2003: The game has not been ruined. In fact, this season has been one of the most entertaining, exciting, and unpredictable in recent history. Although the early season was fraught with doomsayers claiming that Armageddon was upon us because of the distance that some players were hitting the golf ball, the standards in place continue to prevail. As for the "old, classic courses" becoming obsolete, the quartet of Augusta National, Olympia Fields, Royal St. George's, and Oak Hill withstood the test of time and technology once again, as reflected in the facts below: Of the 560 players who participated in the four major championships this year, 15 (less than 3%) finished under par. The aggregate score of the four major champions in relation to par was 20 under. No player shot 4 sub-70 rounds at any of the major championships. Traditionalists also can take solace in the fact that scoring records have not been broken on a weekly basis as originally feared.

With eight players currently averaging more than 300 yards per drive on the PGA Tour, and a total of 60 players averaging more than 290 yards, there is no doubt that players are hitting the golf ball farther. Is it causing the game irreparable harm? Are the "bombers" the only players winning tournaments? Not by a long shot: Of the eight players averaging more than 300 yards in driving distance, three have combined for five wins on the PGA Tour this year. Another three are outside the top 145 on the PGA Tour money list. With the combination of players' improving physical fitness and strength, course conditions, and equipment technology, the professional game continues to grow and prosper because the rules in place more than adequately control technological influence. As evidenced by the performances at the four major championships and at PGA Tour events throughout the year, the depth and breadth of the weekly fields and the talent of rookies and veterans alike should enjoy uppercase recognition.

After all, as the PGA Tour's tagline says, "These guys are good!"

Robert Cupp, golf course architect, on his Highlander golf ball

In 1996, you could build a golf course on 200 acres of land. In 2001, you needed 250 acres to build longer courses to accommodate the new balls and clubs. As we proceed in our industry, we keep wishing that we could find a way to build less-expensive golf courses and reduce their costs and bring more players into the game. Instead, we are buying more land, irrigating larger areas, and planting more grass, all of which costs more and more money. Furthermore, many players are rejecting the $100 a round concept, which I thought was a failed one all along. So we're having trouble making ends meet and discouraging people from playing or taking up golf at the same time.

I own a 1903-built golf course in Birmingham, Alabama, called Highland Park, that isn't even 6000 yards long. It costs $30 to play, sometimes less, and it's overrun with players. For the course's centennial, which is coming up, I am going to produce a 70% golf ball that flies 70% of today's balls' distance, that I'll call the Highlander ball. This will make Highland Park into a virtual 7000-yard golf course, because of the limited flight distance of the ball, and it will bring golfers to that course who otherwise wouldn't come because they felt it played too short. In other words, we're creating the same level of challenge to the golfer on half the area of land, and I'm hoping this project will get national attention and make people think about alternatives to longer and longer golf balls and longer and longer golf courses.

The Contemporary Golf Course Architect: Artist, Programmer, and Engineer

A golf course represents a collaborative work of art between Mother Nature and a golf course architect. For example, when Jack Neville, who designed Pebble Beach in the early part of the last century, first laid eyes on that storied site alongside Monterey Bay in Northern California, he said that nature had already laid out the holes for him. Nature, Neville said, had intended that land to be nothing else than a golf course. The writer Robert Louis Stevenson agreed, as he dubbed Pebble Beach "the most felicitous meeting of land and sea in creation." Golf course architects today still strive to create that same balance with nature that Pebble Beach exemplifies to perfection.

Like their building and landscape architect counterparts, golf course architects/designers work in an interdisciplinary way. They are part artist, part engineer, part environmentalist, and part contractor, with a bit of the businessperson, accountant, and economist thrown in for good measure. Like people working in a myriad of professions, golf course architects rely heavily on their personal computers. Steve Timm calls his computer his "executive assistant": He uses it to organize and file his notes, correspondences, and drawings and plans; to keep track of appointments; to do his billing; and to connect via email to his colleagues and customers.

Mostly, the computer works as an extension of architects' imagination and intellect, making their course designs dramatically easier and faster to produce. Photoshop, for example, enables architects to superimpose hand drawings of a course in progress over scanned digital photographs of the course's site. With a bit of airbrushing and hand coloring, the resulting image presents a believable illusion of the completely finished course. Architects present these photo merges as sales and marketing tools, which give prospective clients a sneak preview as to how a facility will look even before the construction crew breaks ground.

Pebble Beach

The architect also uses the computer to overlay contour drawings of a proposed finished golf course on drawings of the land's existing topography. With the push of a button, a computer program calculates cut-and-fill quantities, or the amount and location of the dirt the builders will remove and relocate to different places while constructing the golf course. In years past, architects had to measure the difference between their contour drawings and the original landscape and then mathematically calculate how much dirt to move to make their course. By taking over the job, the computer not only saves a tremendous amount of time (and, therefore, the client's money), it also produces extremely accurate measurements and easy-to-read plans for the contractor on the bulldozer to follow when sculpting the course.

Computer-aided design (CAD) has become pervasive in the golf course design business, although different firms use it to varying degrees, depending on their size. For example, Jack Nicklaus (who is not only the greatest golfer who has ever lived, but a world-class golf course architect) and his Nicklaus Design firm have become an industry leader in the use of computers to create golf courses.

Bobby Root, Manager of Technical Applications for Nicklaus Design, LLC, on the uses of T2Green International V8.0 computer-aided engineering software for use in golf course design

In 1994, Nicklaus Design contracted me to develop some CAD tools specifically for use in the golf course design process, which, at that time, did not exist. What initially started as productivity tools for drafting plans have evolved into T2Green International V8.0, a truly unique application that Jack Nicklaus and our team of design associates use in designing our golf courses. The software is proprietary to Nicklaus Design. I'd like to highlight some of its features, because it represents the cutting edge in terms of technology in the field of golf course architecture or design.

T2Green has unique productivity tools for automating the placement of nearly every feature pertaining to golf course design in three dimensions. This includes the golf corridors and centerlines, fairways, tees, contours, greens, bunkers, trees, rocks, native and rough areas, waste areas, clearing dimensions, cart paths, lakes, streams, bridges, retaining walls, spot elevations, and so on—all of which are placed into the digital design plan on the computer.

There are obviously others applying CAD technology in their design work, but here are a few key points that set T2Green apart: One, it can produce a golf course routing in roughly 30 minutes! This process once took hours, if not days. Second, the tool for placing the corridor and centerline gives the designer precise yardages as he overlays the golf course onto the existing topography, and creates a running scorecard that can be generated automatically. As design features are added to the course, T2Green keeps a database of all the materials applied to the design. For example, this could be all the square footage or meters of grasses for tees, fairways, and greens. Linear feet of cart path (with or without curbs), cubic yards or meters of bunker sand, and landscape quantities for vegetation are just a few of the elements we can include using T2Green.

Each design plan is generated automatically and is intrinsic to the software. At any point in the design process, the production assistant can generate hole and green details for field books for Jack's review onsite, or any of the design construction documents such as strategy, contours, clearings, cut-and-fill plans, and so on. All this is done at the touch of a button, and automatically comes with the drawing title blocks, designer logos, indicated yardages, and quantitative information about the course.

These features enable us to produce an entire set of plans in an amount of time that is unheard of in this industry. Our software enables 3 production assistants to complete nearly 100 prospect golf course routings and as many as 40 projects or courses per year, far more than they could without T2Green.

We also integrate our earthwork calculation software into the program. We can generate a cut-and-fill plan, which provides the designer with a visual thermal image of where the material on the site can be allocated. This helps minimize the unnecessary addition or removal of excess dirt from the site. We also have tools to assist in designing the residential master plan and general landscaping layout of a golf course development.

The T2Green technology in action

Another tool we have developed and use with T2Green is called a "photo merge." A photo merge is when we take a digital picture in the field of the actual cleared golf course site, say, looking from the proposed tee box toward the green. We then overlay a detailed 3D digital model of the proposed golf hole, done from the same perspective as the photo, over the digital photograph.

But because they enable a photo-realistic image to be created before construction begins, they can be used with great success by clients in their preliminary marketing and promotion of their project.

Although photo merges are quite common in the industry, I would like to explain how we at Nicklaus Design are taking it into the next generation. A photo merge limits you to a static

photographic image, which includes the future holes surrounding vegetation and terrain. We are now utilizing software that enables us to create images of a course or a hole's natural surrounding terrain from various sources to simulate a very realistic rendering of the entire ecosystem of the golf course we are designing, wherever in the world it happens to be. Rather than being constrained by the fixed perspective of a static photo, we can now dynamically render a model from anywhere on the proposed site and define a camera path that captures and creates automated fly-throughs of our proposed designs.

This technology is ever evolving, and in the next two years we envision being able to post photo-realistic panoramic videos to a web site for prospective homeowners or lot buyers, where they can browse specific lots anywhere on the golf course and tour our designs before construction even begins.

The Club at Longview in Charlotte, North Carolina, designed by Jack Nicklaus

James F. Moore, USGA, Director, Green Section Construction Education Program, on technology and tree removal

One of the hardest things for members at a golf club or course to agree to do is to cut down one of their course's trees. Nobody wants to cut down trees, and I don't want to either. But we have so many great old courses that have been ruined by excessive tree planting, because too many trees create too much shade, which doesn't let the grass grow properly. In other words, we still need photosynthesis to make the grass grow. Oakmont Country Club in Pennsylvania, where we played the 2003 U.S. Amateur, is a good example. They removed hundreds of trees to restore the course recently, with absolutely gorgeous results.

There's a company today called Arborcom that goes to a green site and locates the green by global-positioning satellite (GPS), and then identifies by number and species every single tree that surrounds that green. Then they input that information into the computer. Not only that, they also input each tree's growth rates and habits and the exact path of the sun for each day of the year as it passes over the green. What they also can do is simulate the sun's path on any day of the year, and the computer adds up how much sun the green gets on that day, or at any hour of that day. We know that most species of grasses need eight hours of direct sunlight a day for them to flourish.

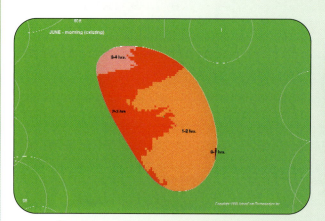 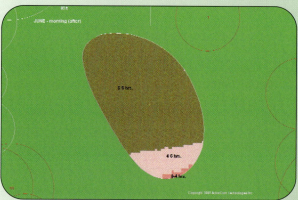

The Arborcom technology with before and after images of the green when trees are removed

As an agronomist, when you go out to a site and look at a green that's weak and you make a recommendation to remove a 150–year-old tree, you have to be 100% certain that removing that tree is going to fix the grass-growing problem. Until now, you haven't been able to do it, because all you could do was guess.

You can say to the computer, "Let's see what the sunlight conditions at this green look like on May 24," and then run a simulation that measures all the areas of the green to see how much sunlight they get on that day. What you might find is that the back-left part of the green is only getting two hours of sunlight. So you go back to the computer and, without pulling out a chainsaw, digitally remove trees number 22, 74, and 73, and rerun the simulation. Now the computer tells us that removing these trees has taken us to three hours of sunlight on that area of the green, so we say, "Let's put tree number 73 back in and take out 75." We keep doing this until we make sure we have the light we need for the green to grow well again before we ever remove any actual trees.

The other thing that this program lets you do is simulate the aging of the trees. What happens at a lot of courses is that the older members say to us, "You know, we never used to have problems with this green 20 years ago." (They're obviously forgetting that the trees were 50 feet shorter then.) Because we know the species of the trees, we can say, "Okay, let's run a simulation that lets us subtract 20 years of age from them and see what kind of light we had then." On newer greens you can simulate what the sunlight conditions will be like 10 years from now.

Even though this process is expensive, what's so nice about it is that removing the trees digitally first gives club members confidence that they are only removing whichever tree needs to be removed and that it's going to work. Consequently, they are more likely to go forward with the tree-removal project, which improves the health of the greens.

(For another approach to solving an excessive green-side shade problem, see Riviera Country Club's superintendent Matt Morton's sidebar in "The Golf Course Superintendent: A Short Course in Course Care.")

Web Tip

For a fascinating and in-depth look at the golf course construction issues, visit the USGA's construction link at their web site: www.usga.org/green/coned.

Robert Cupp, golf course architect, on the golf course as a work of art and engineering

A golf course is a work of art for a very simple reason: People look at it. If you hang something on a wall in a frame, whether it's good, bad, or indifferent, people will look at it and discuss it as a work of art. It's good that golf course architects today are taking more and more of an interest in the beauty of their designs. Although not many of them have been trained in the arts, they are learning to do the little tricks that make courses beautiful. These include things such as leading the golfer's eye through the course's contours and edges as they cross the horizon lines. It has to do with understanding the values of shadows, contrast, and texture and how to compose all three with the movement of the land to make a visually interesting course. The younger architects realize that to be competitive, they have to be able to add beauty to their designs.

A golf course has three primary elements. First is its strategic aspects, which do not have to do with the shapes of the bunkers or the size of the greens, but with the quality of the shots, or the "walk" that the courses ask golfers to take. The best courses engage a golfer's thinking and imagination on each shot. Next comes a course's technology or conditioning aspects. What kind of grass is on the course? How is it maintained? Does it provide a good surface to hit from and putt on? How good is the sand in the bunkers? What is the quality of the grass? And of course, there are the aesthetic issues of the golf course, which have to do with the sheet of clothes with which the architect dresses each hole.

People usually focus on one of these three elements when they say that they like or dislike a golf course. Rarely will a person even pay attention to all three elements on the courses he or she plays—although, of course, that would be ideal. Not many Tour pros even do that.

Golf Course Agronomy: Leaves of (Genetically Engineered) Grass

Ben Hogan once said that the secret of golf lies in the dirt: "Go and dig the secret out of the dirt." By that he meant practice. Although Hogan saw golf as belonging to the practical intelligence (hence, the importance of practice), the science of golf's dirt and the grasses that grow from it comes from the more theoretically oriented labs and classrooms of colleges and universities. Indeed, academic institutions of higher learning in America—

including the University of Georgia, Rutgers, Penn State, Florida, Texas A & M, Nebraska, Michigan State, and others—have helped to elevate golf course agronomy into a dynamic profession that attracts today's brightest science students, whether they play golf or not.

The USGA continues to spearhead agronomic research, and has, since 1983, funded more than 215 university-based projects at a cost of $21 million. This nonprofit research program provides funding opportunities to university faculty interested in conducting golf-related environmental and turf grass management projects and continues to improve golf courses' playing conditions and the enjoyment of the game. Other golf organizations, such as the Golf Course Superintendents Association of America (GCSAA) have helped raise millions of dollars for turf research as well. These monies were funneled to various breeding programs that incorporated the latest technology to produce enhanced turf varieties to be used on golf course greens, tees, and fairways.

The result has been 20+ new creeping bent grass varieties, three Ultra-Dwarf Bermuda grass varieties, numerous new Bermuda grass and zoysia fairway turf varieties, and scores of new bluegrasses and ryegrasses. Most use 50% less water; offer more disease tolerance; offer better resistance to salt, heat, cold, and insects; and offer great improvements in surface texture, and ultimately superior playing surfaces for golf (see sidebar). New organic fertilizers, and slow-release fertilizers, gradually feed nutrients into the soil, which both reduce their runoff into the environment and produce healthier plants. Golf course superintendents can cut this healthier grass to lower heights on greens and fairways during a professional tournament for faster playing conditions as well as let it grow more thickly in the rough to further challenge the Tour pros.

Although not every golf course will benefit directly from this kind of research, all do feel the effects of the new environmental standards that they must meet (see sidebar).

For example, 20 years ago, golf course owners didn't have to do environmental-impact assessments. They didn't have to come up with pest-management plans or negotiate with environmental agencies about which pesticides they planned to use on their courses. In fact, one of the primary challenges facing turf grass managers (that is, agronomists and golf course superintendents) today is their need to familiarize themselves with the best management practices for so many new varieties of plants. Until the late 1980s, golf course superintendents had only 3 to 4 primary creeping bent grass varieties to choose from…by the mid-1990s they had more than 5 times that number. What's more, in the past, if someone owned a swamp, they could fill it in and build a golf course. Now people can't touch the wetlands or build courses next to existing streams without convincingly proving they will not harm the water in any way.

In other words, the public and the government alike used to look at golf courses like small mom-and-pop farm operations, where the owners could do to the land whatever they wanted with impunity. Today, developers have to ask themselves if they have the grasses and technology to meet today's strict governmentally enforced environmental requirements? If they don't, they have to meet them before they can realize their vision of building a golf course.

Dr. Michael Kenna, Director of Green Section Research, USGA, on breeding and engineering new grasses for golf courses

Most of the research at the USGA not only addresses issues facing golf courses today, it also anticipates the problems they will encounter in the future. For example, most of our efforts presently concentrate on the quality of and quantity of water we need to use on golf courses, and other environmental issues, and these will certainly continue to challenge us as time goes on. There has been a trend for golf courses to change from potable, or drinking-quality, water for their irrigation to recycled or treated sewage wastewater, which, of course, meets EPA rules and guidelines for quality. Courses in Southern California, Arizona, Texas, and Florida have led the way, and many now successfully use tertiary or treated wastewater on their courses. The turf and soil act as a natural filter for this water and clean it as it seeps downward before finding its way back into streams that feed the aquifers.

The value, of course, in using treated wastewater is that it conserves drinking water. To use this water on courses, however, we have had to develop hybrid grasses that will thrive when irrigated with this lower-quality of water, which has a high salt content. So a lot of agronomists' work

Photo Copyright
John and Jeannine
Henebry Photography

The Colliers Reserve course with Seashore Paspalum grass

today has involved breeding grasses that resist or tolerate salinity, or salt. A good example is the Seashore Paspalum grasses, an inherently salt-tolerant species of grass that has been bred for use on golf courses in hot and humid regions close to the ocean. Colliers Reserve, a course in the Naples, Florida area, actually waters their Seashore Paspalum with briny, estuary water, which is almost as salty as seawater.

We continue to support research toward genetically engineering grasses such as creeping bent grass for a greater tolerance to salt, but we really haven't succeeded in this area as of yet.

We have, however, made advances by conventionally breeding creeping bent grasses that require less water, while at the same time working to expand the geographic range where we can use Bermuda grass. Bermuda grass is typically used today on courses throughout the South and on those as far north as Oklahoma and Maryland.

If we can develop Bermuda that thrives in cooler climates farther north, this will also preserve water, because such a grass requires less water than the perennial ryegrass or bent grass typically used there. Where the winters are too cold for our current Bermuda grass varieties, we will have a golf course grass that will require less water during the summertime and provide a better playing surface than perennial ryegrass. We're working on breeding winter hardiness into these new strains of Bermuda, so it will survive the cold months in the northern regions.

Where genetically engineering has shown some success is in developing grasses that resist insects and other pests. In fact, scientists at Rutgers University in New Jersey, along with the Scotts Company, have succeeded in genetically engineering a herbicide-resistant bent grass. The herbicides will kill the weeds in this grass, but won't harm the grass itself. That product is presently going through the arduous deregulation process with the USDA (United States Department of Agriculture).

Robert Cupp, golf course architect, on transplanted golf courses

The technology in agronomy and golf course construction today has allowed the golf course architect and superintendent to grow healthy grasses in places now where it couldn't grow before. It's all because golf keeps growing and people want to play. Subsequently, we're trying to find homes for courses where we don't have to pay a screaming fortune for the land.

For example, one result of the technology is that people can play on and appreciate again the original type of golf course—in other words, the links and links-style course. One of the defining traits of a true links course is that it is situated next to the ocean. This means it has wonderful sandy soil that drains vertically straight downward. With the technology in irrigation systems that we have today, we can build those very busy rolling, dipping, naturally undulating shapes that characterize a links course, and still drain them, even though they don't have that wonderful sandy soil that drains so naturally and well. Jack Nicklaus's course at Grand Cypress has many greens based on those at St. Andrews, and they look and play like a links course in Orlando, Florida, with no ocean in sight. What Pete Dye is doing with the links style

of architecture in Kholer, Wisconsin, is absolutely spectacular. I did one in Lake Orion, Michigan, called Indianwood, that has hosted two U.S. Women's Opens, and that is literally a links course without the ocean.

Technology also has enabled us to build wonderful courses in the desert where no grass grows naturally. We have learned how to manage the desert's flora and fauna, so we can now make desert-landscaped courses that aren't contrived looking, meaning they do not present wall-to-wall carpets of green grass as they did in decades past. What we see instead is what we call desert landscaping that presents gradual transitions between the lush green grass of the golf course and the obviously more dry and bare desert itself, and they are often quite spectacular to look at. And, again, we couldn't do this without today's advances in agronomy and irrigation, because we have developed grasses that don't require as much water. So they may not have the plush bright-green appearance of grasses there in the past that required a lot of watering, but are wonderful surfaces from which to hit golf shots and they're better for the environment.

Robert Cupp's Indianwood

Billy Fuller, agronomist/design associate at (Robert) Cupp Design, on why golf courses are good for the environment

I cannot overemphasize how positive the impact of golf is on the environment, although many golfers, and even people working in the golf industry, don't fully understand why. Let's forget that I am in the golf industry for a moment and focus on these facts:

- Turf grasses are the best "natural" filter known to man.

- The average golf course offers enough carbon dioxide/oxygen exchange for 150,000+ humans daily.

- Studies performed by the U.S. Environmental Protection Division in the mid-1980s on four golf courses in the Cape Cod area proved that water quality was more pure under the golf course than either upstream or downstream of the golf course.

- Wildlife populations are greatly enhanced by "edge habitat," and skilled and environmentally sensitive golf course architects build their courses with multiple edges throughout. Examples are tree line edges, lake and pond edges, stream/creek edges, edges created by changes in turf grass varieties, edges created by varying height of cut, and so on. These edge habitats establish and provide a range of environments in which different wildlife species thrive.

The impact of technology on agronomics and maintenance is, however, a two-edged sword. On the one hand we have greatly enhanced turf conditions that offer immaculate fairways and smoother, faster, and more blemish-free greens. On the other hand today's golfers in North America are spoiled to the point that their expectations at some clubs far override common sense. They forget that golf course superintendents cannot control conditions such as drought or floods brought on by Mother Nature, and that no turf grass is immune to these sometimes extreme pressures. For these reasons, I believe the golf agronomist and golf course superintendent must do more to educate and inform the players at their clubs and courses to have more reasonable expectations about the conditions of their golf courses from day to day and season to season.

Golf Course Construction: Michelangelo Under Par

Anyone who has ever built a house or an extension knows the hopes and the fears, as well as the anxieties and creative pleasures, of working with a contractor. Building a golf course is no different. A developer or owner may have the initial idea for a course, and the golf course architect will flesh out that vision and draw the plans for it, but a golf course doesn't amount to a hill of beans until the builder puts it into the ground correctly. After the architect completes his or her last cut-and-fill calculations, and the agronomist selects the types of grasses, plants, and trees for the course, the contractor rides in on steeds of powerful bulldozers ready to break ground. In fact, the men and women (called "shapers") who mount these machines and sculpt the earth into golf courses represent the unsung Michelangelos of the golf world. Unknown and, perhaps, underappreciated by the general golfing public, the golf course developers and architects who hire them to work their magic value them like gold.

Builders today turn to the heavens for a little help to map the ground in the initial stages of building a golf course. More specifically, they use GPS systems to accurately landmark the placement of vital golf course apparatuses. These include irrigation heads and valves, water pipes, and the specific coordinates for their cut-and-fill land-moving operations. They also look under the ground with infrared technology to analyze the nutrient makeup and quality of the soil, which tells them how much fertilizer they will need and where to administer it on the course.

The contractor usually begins by imagining the completed golf course, because although the importance of getting water on to the course to irrigate it seems obvious, it is equally important to get it off the course so that the same grass can breathe. To this end, drainage pipe manufacturer Hancor recently introduced a dual-wall, high-density resin-based product called EcoFirst, made from HDPE, a high-density resin. Constructed from environmentally friendly recycled content, the product features a corrugated outside for strength and a smooth inner wall for increased water flow. Many new and renovated golf courses (such as San Francisco's Harding Park) have turned to EcoFirst, which has earned the approval of environmental watchdog organizations such as the Audubon Society and Ducks Unlimited, a large land conservationist group.

The EcoFirst Technology from Hancor

Builders also use a relatively new and simple drainage pipe called a flat pipe, whose width far exceeds its height. Laid flat on the ground under the surface of a fairway or green, a flat pipe presents more surface area for water to flow into and through than its circular counterpart, which makes it particularly useful for green and tee areas.

Irrigation (as opposed to drainage) pipes also last longer today thanks to computer-regulated pumping systems that reduce wear and tear, or the "hammer effect," on the pipes. Old-style pumps pushed water through pipes at as much as 1200 to 1500 gallons per minute. Today's computer-controlled, variable-frequency technology gradually revs the pumps to supply whatever demand is needed and/or programmed, and this greatly reduces the amount of stress placed on the piping system. Many developers have turned to a low-cost and relatively new style of green construction called the California Green, which under certain conditions provides superb drainage and a quality putting surface (see sidebar).

Previous chapters discussed how the new golf balls and drivers have forced architects to design longer and longer courses. Similarly, today's green speeds have increased dramatically as a result of new and superior turf grasses and the improved maintenance equipment and technologies that keep them alive at incredibly short lengths. One company has developed a fascinating technology to retrofit classic greens to better accommodate today's lightning fast grasses (see sidebar).

As in all areas of golf, new and often mind-boggling technologies continually surface both to make the contractors' job easier and to improve the quality of their 18-hole products. In fact the next wave of jaw-dropping golf technology may very well come from the building segment of the industry. For example, manufacturers of heavy earth-moving equipment have begun research and development on Buck Rogers-like automated bulldozers equipped with lasers and GPS technology. Someday soon, the blades of these machines might just follow a computerized program to carve a golf course out of the ground all on their own. Although some might argue that such a procedure would do little more than rob the job, and the final product (that is, the golf course), of its romance, the future, in the words of the legendary comedian Mort Sahl, lies ahead. Even a million cubic yards of earth waiting to become a golf course probably won't stop it.

James F. Moore, USGA, Director Green Section Construction education program, on technology and rebuilding greens

Putting greens have gotten progressively faster in recent years as we have bred better grasses that have allowed us to lower the cuts on the greens. In fact, if you look at the old films of golfers putting, they show players basically taking a half swing to get the ball rolling. The old putters, in fact, all had considerable amounts of loft built in to them to get the ball rolling over the greens. Now until four or five years ago, we had what I call a built-in fuse in the grasses we were using on greens, meaning we could only keep them at a low cut for a limited amount of time before they would start to fail. So you could take a Pentcross bent grass and mow it down real low during a tournament week, say to an eighth of an inch, so it rolled very fast. As soon as the tournament week ended, however, you had to let it grow longer again. That has changed today, with the new heartier breeds of grasses we have that can be cut low and kept low, in many parts of the country, all summer or playing season long. Of course, not every course has the money, equipment, expertise, or need to do this; but if everything is right, it's much more feasible to have greens rolling at 10 feet on the Stimp Meter all season long. This has affected many of the greens on the old classic courses, such as those designed by Tillinghast, Donald Ross, Alister MacKenzie, and others. These architects and others designed their greens with a lot of sharp and heavy slopes and contours (up to 6% or 7%), without concern, because the slower grasses wouldn't allow the ball to roll too quickly on them. Today, with the heights of grasses that we have, the ball hits a slope of 3% or more and it will roll back toward the golfer, sometimes right off the green. So the real problem becomes finding enough of what we call "cuppable" surface areas on these greens to place the hole in a fair manner.

In the past, you needed a transit and rod to survey the green to determine the areas of slope in small intervals or "intersects" on a green. This would take forever, and hardly anyone does it anymore. A company called GolfTech developed a new technology that utilizes both laser and GPS and a robotic surveying piece of equipment that shoots every intersect on a green within an hour's time. The robot follows a person walking across the green and reads and records the green's exact slopes. This data is then put into a CAD system, which shows each intersect and tells whether it has 1%, 2%, 3% of slope, or whatever the number. So when the members at a classic course want to rebuild their greens, the architect doing the job can use this technology to build new greens that are modeled after and follow the general contours and slopes of the old greens, only not so severely. The result is greens that maintain their beautiful artistic and classic shaping, but still have enough cuppable areas for a variety of pin placements, which preserves the integrity of the course.

Preparing the GolfTech technology for action

Dr. Michael Hurdzan, golf course architect, on the California Green

I developed the California Green based on research done 20 years ago at the University of California, Davis, which was based on the idea that the best medium for growing putting green turf was pure sand, with nothing added to it. The research showed that if you gave exclusively sand-based greens the water and nutrients they needed, they would grow. In recent years, the California Green has gained in popularity around the world and is used as a cost-effective alternative to the USGA level green. Here is a simple explanation of the difference between the two types of greens.

As I said, the California Green is basically putting turf laid over a bed of pure sand that a laboratory analyzes to make sure its particles are basically the same size and that it has excellent drainage properties. In fact, the theory behind the California Green is that water will move through similarly sized particles better and faster than through particles of different sizes. This sand bed of a California Green can extend anywhere from 15 to 18 feet downward from the green's surface, where perforated pipes collect the water and transport it away.

The USGA level green starts with a layer of gravel on top of which it layers sand with organic matter, such as peat moss, mixed in with it. The USGA greens drain faster initially, and then retain water in the root zones, whereas the California Green drains more slowly, continually, and from my perspective, more thoroughly. USGA greens tend to do better with a high quality of water applied to them (for instance, drinking-quality ground water), whereas California Greens do very well when irrigated with effluent or treated wastewater.

The advantages of the California Green from my point of view are that they are far less expensive to build, and if constructed correctly, provide as good a putting surface as any other. Those who object to the California Green believe that greens need that gravel layer, and the amendment or organic matter added to the sand layer, to conserve the water for the plants' root system as the water drains out of the greens.

The Golf Course Superintendent: A Short Course in Course Care

Today's golf course superintendents are highly educated, skilled, and talented individuals, who work long hours so that golfers can more thoroughly enjoy their courses. Almost all have academic backgrounds in plant physiology and soil science. Many have graduated from one of the several golf course management programs offered at colleges and universities. Their charge is to care for the living and breathing organism called a golf course. To do so, they must stay abreast with the technology of their dynamic profession that changes perhaps even more quickly than the golf club and ball businesses. Just the technology of today's irrigation systems, designed to minimize water use, boggles the mind for its sophistication and capacity for pinpoint accuracy.

For example, 20 years ago golf courses used electromechanical timing devices that turned on blocks of up to 10 sprinklers at a time. Today superintendents program centralized computer-controlled systems with separate sprinkler heads that target areas on the course that need watering. In arid regions such as Las Vegas, Southern California, Phoenix, and Dallas, ground sensors balance calculations of dew formation, soil and grass evaporation levels, and ground water deposited by the rain, and then automatically program sprinklers to ration out the water the course needs to stay healthy.

A golf course superintendent at work

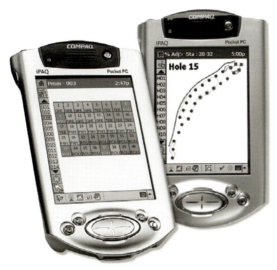

The Toro Prism Pocket PC remote irrigation system manager

Photo courtesy of the Toro Company

What's more, the superintendent can turn many of these irrigation systems on and off with a handheld remote-control unit or a cell phone. So, instead of wasting time walking to and manually turning on the sprockets of an individual watering unit, the superintendent can survey the course in a cart and start the sprinklers at the touch of a button. Many of these computerized systems work with a high-tech weather station located on course, which measures humidity, solar radiation, wind speeds, soil moisture, temperature, length of the day, and other data that impacts the course's evaporation rate and level. The station sends this data to a central computer, usually in the superintendent's office, where the superintendent reads it and then decides how much water the course needs and where it needs it.

Today's (and tomorrow's) lawn mowers barely even resemble those of the past. Many courses have switched from the conventional loud and heavy gasoline mowers to silent and lighter electric-powered ones, which reduce both annoying noise pollution and turf damage. Manufacturers are already experimenting with robotic fairway mowers that use GPS to cut the grass with no one operating them.

Superintendents are not just the custodians of their particular golf courses. They also must keep the best interest of the environment as a whole first and foremost in their minds. Therefore, to those ends, they diligently stay abreast with the latest environmentally sound agronomic advances and practices. For example, some tractors today come equipped with GPS that know the exact location of weeds on the course and sensors that signal and administer herbicides to them directly, thus reducing the amount of herbicides used on the course.

The Aurora system from Signature Control Systems uses field-control stations, like this one, that can function as stand-alone controllers or satellites that send data to a central computer.

Some superintendents even use Zeolite and other classes of inorganic amendments originally developed through NASA-funded research to grow crops on the moon and other planets. They spray this material around fertilizers and/or herbicides and pesticides, which forms coats around them, which time-delays their release into the soil at the rate that the plants will use them. Like Zeolite, wetting agents called soil surfactants coat the soil's particles, which allows them to hold more moisture and, again, reduces the amount of water superintendents need to use on their courses.

Golf course superintendents then are a hybrid of the Jacks-of-all-trades, the family doctor, and a midwife. They care for their leafy patients through maladies both serious and small (see sidebar) and nourish them toward a healthy state of independence. They may have to repair a rut dug in the rough by a stuck golf cart in the morning, and then run to an afternoon meeting with agronomists and botanists about eradicating a disease that threatens all the course's Eucalyptus trees. They may prepare their courses into near perfect conditions for a PGA Tournament one week, and then struggle to live up to their own high standards to please their membership for the rest of the year. Just like playing the game well, the superintendent's job is difficult and requires talent. In fact, it is not so much a job as a calling, which superintendents answer with modesty, enthusiasm, and joy.

Matt Morton: Head Superintendent, Riviera Country Club, on techno-caring for the famous 6th green, and other greens

Some of my favorite golf courses are St. Andrews and the other courses in and around Europe. I love watching the British Open because the courses look so natural. Yet I know that even these courses are making technological changes. American courses, especially ones (such as Riviera) that host a PGA Tour event (ours is the Nissan Open each February), have a perfectly manicured look with perfect playing conditions. Our members have come to expect that week after week, and we pretty much provide it, because that's what, in part, they feel they are paying for. But what interests me is leading-edge technology that improves my course but doesn't rob it of its natural beauty and look.

We just installed a new subsurface aeration unit at our second green, which is a USGA green with good drainage pipes underneath it. This system attaches to those drains and uses an oxygen sensor that makes sure the roots of the grass are receiving enough oxygen. When they aren't, the sensor kicks on a pump that sucks air through the green's surface, which moves oxygen throughout the cavity of the green for the roots. We are currently monitoring the root zone weekly where we think there will be an increase of root mass. These systems could be installed in all the greens at the course to help move oxygen through our compacted soil and help to maintain quality putting surfaces. The system will also suck water out of the green's cavity, which means it doubles as a drainage aid as well. We've buried the machinery under the ground off to the side of the green, so players—and, of course, fans during the Nissan Open—can't see it.

Before I talk about the 6th green, whose sand trap in its middle makes it one of the most unusual and famous greens in all of golf, I want to make the point that every green on a golf course has its own microclimate that requires individualized care. Some are shaded, some are not; some have trees, some don't; some are big with a lot of slope, others are flat. The 6th green at Riviera has historically been a troubled green. Superintendents in the past have had nightmares about the health of this particular green more so than any other one on the course.

First, it sits right beneath a tall canyon wall, so it doesn't see full sun before 10 a.m., particularly in the winter, and greens need morning sunshine. Next, greens need air movement—a nice light breeze is healthy for grass—and that green sits in a corner where the air is often stagnant. The third thing is that the soil temperature below that green is colder than any other green on the course—again, because of the reduced sunlight.

Riviera rebuilt the 6th green in 1996 after the PGA Championship, which was played here, and the superintendent at that time put in a heating

system underneath the green's mix. A hot-water pump off to the side of the green feeds the system's small pipes, which run through the entire green. The system's thermostat gave the superintendent control of the soil temperature, so the preferable soil temperature could be provided to maintain healthier turf.

Next we transported in fans around the green, which we use on humid and still days to create needed air circulation. We have the ability to remove them and place them at another green if so needed.

In 1998, we installed four grow lights, the kind used in greenhouses, around the green to see whether we could artificially counter the problem of insufficient sunlight, especially in the winter. We try to mimic the sun, so we turn them on at about 6:30 in the morning in the winter and they click off at the end of the day. In the summer, they turn off at about 11 a.m., when the sun takes over. We have the ability to remove them before major events so that they do not affect play. All of these technologies are helpful tools that are becoming common in the industry and help us at Riviera keep our signature 6th green and others in top-quality condition.

Riviera's 6th green

The Golf Car: GPS, Beam Me Down

Long before today's long-hitting drivers and golf balls, the popularization of the motorized golf car began to change the very basic shape of golf courses. Prior to the cars, architects designed courses so that golfers could leisurely shuffle a few steps from one green to the next tee. Today's 7000 yard-plus layouts give players little choice but to jump in their golf cars for a car-path-thrilled race over hill and dale, often through dense housing developments and short tunnels, before resuming their games again. Thankfully, new golf cars from manufacturers such as Club Car, E-Z-Go, and Yamaha make the trip more enjoyable and comfortable than ever. What's more, cars equipped with GPS golf course management systems and monitors (from companies such as ParView, ProLink, ProShot, and UpLink) show golfers vital information for their rounds on display monitors mounted on the cars. This includes their exact yardages to the pin (which satellites triangulate from outer space), positions of sand bunkers and lakes, and other strategic data. Among myriad other features, these cars present a food and beverage menu programmed to pop up on their car's screen so that golfers can order lunch, which will be waiting for them at the turn.

Much of the technology found in today's golf cars comes directly from the automotive industry. The better cars (and they are *cars*, whereas *carts* are something people push), for example, have precision rack-and-pinion steering and comfortable shock-absorption systems. Companies manufacture their car panels with lightweight, flexible, and durable injection-molded thermoplastic elastomer materials, similar to those used on Saturn, and other, automobiles. Golfers living in gated and residential golf communities, such as PGA West in La Quinta, California, can order their privately owned golf cars with head and tail lights, horns, and other accessories, which allow them to drive to and from the course on public roads.

E-Z-Go's Red Freedom

Although federal regulations prohibits golf cars from riding faster than 15 mph on the course and 19 mph off it, improved technology has made both gas and electric cars more efficient than ever. In years past, for instance, cars had one set speed and one acceleration rate. They also had one regenerative breaking (meaning the motor starts to act like a brake when the foot is removed from the go pedal) rate. Club Car's recently introduced IQ technology in their electric cars, which allows managers to custom adjust all three of these variables in ways best suited to the course's terrain and conditions. (For example, mountain courses would opt for rather strong brake settings.) Car barn workers can now computer program their electric cars at night to begin recharging their batteries during the least-expensive electricity rate hours. Gas golf cars burn fuel more efficiently than ever, and overall weights of both gas and electric golf cars have started to come down. This means less harmful impact on both the course and the environment. The split between electric and gas golf cars at American courses stands at 70%/30% in favor of electric.

Club Car and E-Z-Go also offer single seat electric cars for one golfer. Happy to comply with the Americans with Disabilities Act, golf course managers often order a few of these vehicles, generally equipped with hand-operated controls, for injured, ill, or disabled players. Of course standard golf cars extend the playing days of older golfers, who can no longer walk the course. Interestingly, the first golf cars, patented in 1948, were called "Arthritis Specials."

ParView's GPS system

GPS management and communications systems such as those mentioned previously turn a round of golf into something that borders on a high-tech spectacle. These devices seamlessly integrate a host of services, starting with high-resolution, full-color, hole-by-hole graphic overviews of each hole, often replete with 3D renderings of fairways, bunkers, greens, and trees. Again, the GPS system acts like a dutiful satellite-assisted caddy to tell golfers their exact distances to the pin. They also can provide tips from the course's pro on how best to strategically play each hole. (For example, "This hole has a lake to the right at the bottom of the hill that you cannot see from the tee. Favor the left side of the fairway with your drive.") These tips can greatly help golfers playing a course for the first time. Golfers who walk the course can use a small handheld GPS device, such as Sky Golf's SG 2. They also can get accurate yardage readings from one of several small handheld ranger finders (including ones from Nikon and Bushnell) that use lasers rather than GPS for their measurements. All should know, however, that the Rules of Golf prohibit the use of any mechanical yardage provider, so golfers cannot use them in official competition or in rounds posted to determine their handicaps.

A golf car from Yamaha

to every car on the course, telling golfers to drive in immediately. Conversely, someone in medical trouble anywhere on the property can push an emergency button that tells the pro shop to send help right away.

GPS units also can display a company outing's tournament scores on the video monitor. They can beep or flash a warning signal reminding golfers not to drive too close to environmentally sensitive areas or to greens and tee areas, and then automatically shut down the car if a driver fails to comply. They can post display advertisements of businesses large and small, and allow car barn managers to monitor car use so that they can equally spread it out between the entire fleet over time.

Although golf cars have become a ubiquitous part of the American golf scene, they nonetheless remain a topic for debate. Do they rob the game of an essential characteristic—that is to say, walking? The PGA Tour originally argued that they do in their much-publicized legal contest with disabled pro Casey Martin, who sued to use a golf car in competition. Do they not only allow golf course architects to build longer courses often over hilly, mountainous, marshland, or otherwise unwalkable terrain, but also encourage it? Whatever people think, the golf car is here to stay. Perhaps their presence will remind people of walking's lost pleasures, and developers will start building golf courses again where walking or riding become a feasible choice.

Today's GPS systems also provide potentially life-saving two-way text communication between players in their cars and off-course personnel in the pro shop or clubhouse. For example, a worker in the office can send out warnings of an approaching thunderstorm

Colliers Reserve

Photo Copyright John and Jeannine
Henebry Photography

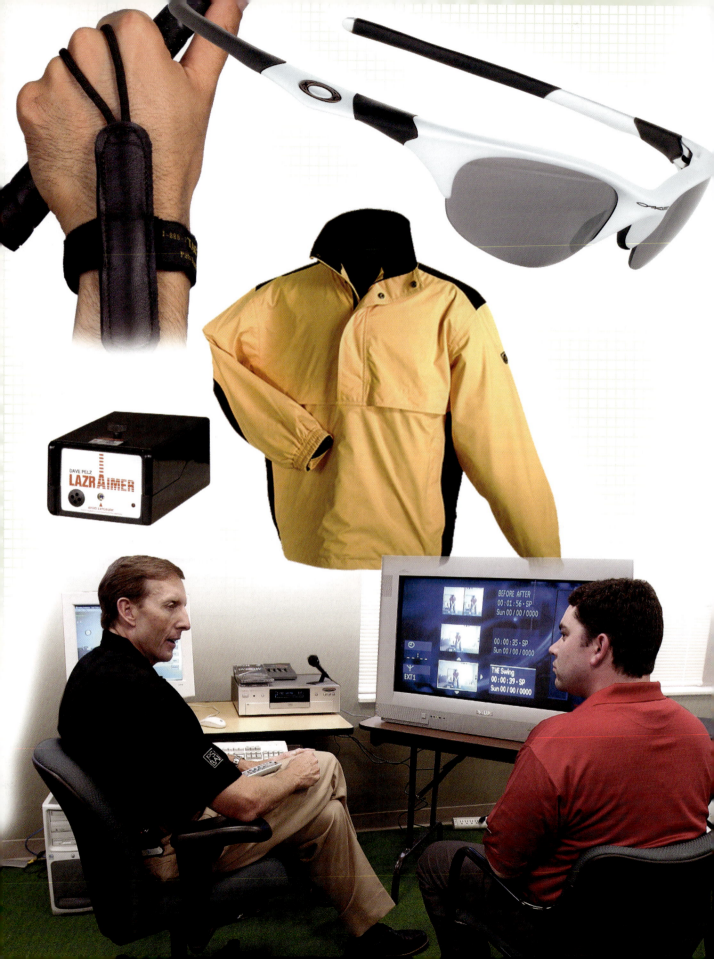

DAVE PELZ
LAZR AIMER
AVOID EXPOSURE

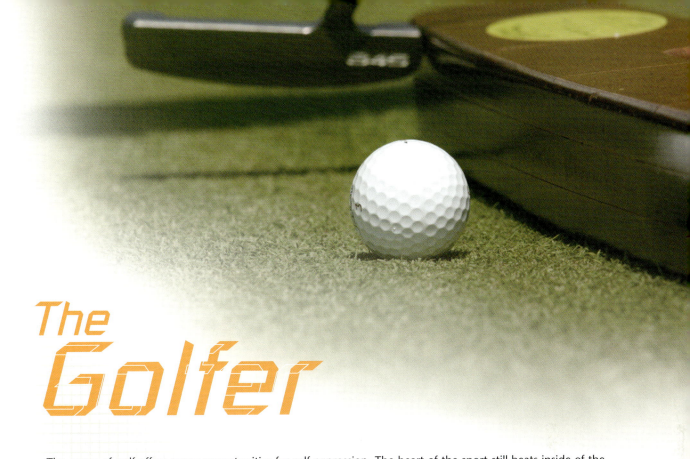

The
Golfer

The game of golf offers many opportunities for self-expression. The heart of the sport still beats inside of the men, women, and children who play it. As long as this is true, golfers will continue to explore new and creative ways to play the game better and enjoy its glories even more.

For example, while every player wants to develop a good golf swing, more and more enlightened golf instructors now realize that sound technique does not mean one formulaic swing that fits all. Rather, they teach from a base of sound fundamentals that maximize their students' strengths and work around their limitations. To help, a host of new training or teaching aids has arrived on the scene. Some assist teachers in analyzing their students' swing flaws, while others allow golfers to work on their own games independently. Of course, golfers have more instructional material available—literally at their fingertips—as the Internet has allowed golfers to take lessons and read instructional material from each corner of the globe. Non-golfers often look at linksters' fanaticism and wonder if people aren't taking a simple game too seriously? Certainly, even the most severely stricken golf nut needs a bit of comic relief once in a while. For diversion without the dread of making a double bogey, golfers can enter the ever more virtual world of golf video games.

Away from the computer, exercise specialists have developed golf-specific training regimens that include strength and aerobic training, and sports psychologists work with golfers on improving their mental acuity. There has been a growing recognition of good nutrition's role in playing better golf, and ophthalmologists and dermatologists have made progress in convincing golfers to protect themselves from the sun. In short, serious golfers, like ancient Greek Olympians, strive for an ideal balance of a healthy body and a sound mind.

While the intersection of the sartorial and the technical may at first seem hard to locate, apparel makers routinely add elements of utility to their new golf lines. Not only does today's golf clothing look good, but it also keeps players dryer in wet weather and cooler on hot days, while allowing freedom for the swing's motion even in pouring rain.

Swing Training Aids

Great golf swings, like Tiger Woods's and Annika Sorenstam's, repeatedly deliver the clubhead squarely into the back of the golf ball. Although the game's best players swing the golf club in one fluid motion, they often build their swings (and often work on them when they go astray) one component part at a time. Swing training aids can greatly assist and expedite this process. Today, golfers and teachers alike can choose from literally hundreds of trainers designed to focus on and improve the individual parts of a golf swing. Teachers often help golfers piece together a complete game by prescribing what might be called "swing trainer cocktails," or several different training aids that help students develop a sound, well-rounded game. Of course, a complete game includes putting, and golfers have many new and interesting putting aids from which to choose.

Swing trainers often present themselves in imaginative shapes, sizes, and modes of functioning. They creatively push, pull, prod, and all but provoke golfers to do one thing: mold a player's golf swing closer to that holy grail of repeatability. Some are as simple as a piece of impact tape that attaches to the clubface and records whether golfers struck the ball squarely or off-center. Others use sophisticated digital video technology so that golfers and their teachers can view and analyze their swings. After familiarizing the golfer/student with the correct bio-mechanical movements of the swing, training aids turn the job back to the golfer, who must translate those mechanics into subjective "feels" or images as quickly as possible.

Digital Video Swing Training Systems: Candid Cameras

Noted teaching pro Jimmy Ballard popularized the use of video in golf instruction after his then student, Curtis Strange, won the U.S. Open in 1988 and 1989. Ballad had identified several key positions and movements in the golf swing, which he believed when mastered guaranteed quality ball striking. He would review his students' swings on tape during a lesson, and then work with them on the aspects of their technique that they needed to improve to bring them closer to an ideal motion. In those days, teachers would just play the videotape back on a television monitor, occasionally hitting the slow motion and stop action buttons and draw lines with magic markers and rulers on top of the student's image to illustrate the content of their lesson. Even so, thanks to video, golf pros or average players could see their swings as their teachers saw them, and this facilitated clearer communication and understanding between the two, which led to faster learning.

Video teaching took a huge step forward in the early 1990s, when a company called Astar replaced the simple analogue video camera with high-tech digital video equipment. This both instantly and almost exponentially expanded the range of the teaching services these newly dubbed digital video analyzers could provide. Today several companies, such as Swing Solutions and V1 (in addition to Astar), lead the

field in digital video analyzer technology. Some of the features these new systems offer include club and strobe-like body tracking, which traces the club and body in motion in individual frames, and split-screen comparison modeling, which allows teachers to pull up a video of a Tour star's swing and juxtapose it against their student's for comparison purposes.

The video camera (or multiple cameras) of today can now capture the swing from different angles. Top systems integrate drawing tools or "telestrators" that, with the tracing of a finger on the monitor screen, enable instructors to easily mark correct plane lines, head movement, impact position, and other basic swing fundamentals right on to the monitor. Although no one expects a high-handicapper to swing like Tiger Woods, teachers can use all of these features to help students see and develop important general fundamentals that all good swings share. What's more, the manufacturers have made all of these systems much easier to operate than in years past.

The Swing Solutions System

websites. The teacher then creates a lesson for that student by using all the video analyzer's graphic technologies mentioned previously. He or she then streams the video back to the student over the Internet.

The Virtually Perfect Golf Learning System, from a company called Strolf, takes video-based learning into the third dimension. The golfer/student dons a pair of 3D virtual-reality glasses, which projects a computer-generated outlined silhouette of a golfer performing a technically correct golf swing. The technology enables student golfers to "step inside" the lines of the computer-engineered character, and then receive instant feedback by seeing in real time just where their swing does and does not sync-up with the model's.

Swing Solutions points toward the future of digital video systems with a newly developed lightweight and portable unit designed for students to easily use themselves on the practice range. After their lesson, students take this monitor to another practice station, where they film themselves hitting balls. They can then compare these new swings with those from the lesson they just received, while their pro's instruction remains fresh in their minds.

Companies such as Astar and Swing Solutions have also taken their technology online. Students can buy motion-capturing software, video their own swings at home or at the driving range, and then transmit them over the Internet to their teachers at their

Should people be weary of focusing on so much hand-eye coordination, they can shift into some *finger*-eye work with one of today's cool golf video games (see sidebar).

Any golfer serious about improving his or her swing must, in the final analysis, develop a conception and image of the swing as one single fluid motion. Video technology, by providing clear, moving pictures of a student's swing, helps golfers do exactly this.

David Leadbetter, world-renowned golf instructor and founder of the David Leadbetter Golf Academy, on video swing analysis systems (www.leadbetter.com)

Using video as a tool is very helpful in teaching golf, and we've been doing just that since the start of the David Leadbetter Golf Academies. With all the new computer software and digital video equipment, you can now do the same things the old VHS cameras did and much, much more. For example, the new systems such as the V1 software we use at my World Headquarters Academy at ChampionsGate, not only enable you to draw lines that illustrate the various swing angles, but to quantify them as well. You can pull up the swing of a great Tour player such as Ernie Els or Nick Price and run it next to the student's swing in a split-screen format for comparison purposes.

One of the other great things we can do now is store our students' swings on a DVD. VHS tapes degrade over time; the DVD recorder enables golfers to enjoy their lessons far longer than with VHS recorded lessons, because DVDs last much longer than VHS tapes. The digital recording will allow students to compare their swings throughout the years as they continue to improve their game. With the added creation of an index picture screen, the DVD recording also allows golfers to quickly and easily find the most important parts of their lesson. No winding or rewinding needed to review your lesson summary. DVD technology also gives players the flexibility to review their lesson on their laptop, on their office PC, or on their home DVD player. At my academy at ChampionsGate, we store all our swings right in the computer itself and back everything up on DVDs. This eliminates having VHS tapes lying all around our teaching facility as they did in years past. Remember, however, that although all this new technology is great, it still doesn't swing the club for the student. It's important to remember that technology is a tool that's only as good as the teacher using it.

Another advantage of today's systems—such as the V1 software—is the ability to do lessons over the Internet. This has become an invaluable tool for me personally, allowing me to keep in touch with not only my players on the PGA European Tour, but players on the U.S. PGA Tour as well. A great number of my players—such as Justin Rose, Aaron Baddeley, Lee Westwood, Ian Poulter, and the like—will send me swings from tournaments as well as from home when they are taking a week off and are unable to make the trip to Orlando to see me in person. This allows me to

David Leadbetter (left) at work with a student

keep a better handle on how the things we are working on are affecting that particular player's swing motion. It really has proven to be a great asset to not only myself and my Tour players, but to golfers in general. We now have several dozen students of our academy who are keeping in touch with their instructors via V1 and the Internet. Coupled with our extensive video and computer equipment, we have also begun using a new ball-launch monitor. It gives us data such as clubhead speed, ball speed, backspin, carry distance, maximum height, vertical launch angle, and horizontal launch angle. It helps us quantify a player's impact data and has proven to be another great teaching aid and clubfitting tool.

Video is now moving into 3D, and we'll soon be able to quantify things such as weight distribution not only at address, but throughout the swing as well. We'll be able to look closely at things such as gains and losses of velocity at different places during the swing and look into why this is happening. All this being said, we have some very, very exciting times ahead in the world of golf instruction!

Jason Frankovitz, Production Assistant, X-Play, TechTV, on Tiger Woods PGA Tour 2004

Tiger Woods PGA Tour 2004, the latest golf title from industry behemoth EA Sports, offers an impressive variety of gameplay and includes several new features not present in the 2003 version.

Realism is a hallmark of EA Sports titles, so many real-world courses are available such as St. Andrews, Spyglass Hill, Pebble Beach, and additions for this year include Kapalua, Pinehurst, and Bethpage Black. All the top-ranked players are here too, with in-game performance modeled after their actual play styles. Among your opponents are John Daly, Adam Scott, Jim Furyk, Colin Montgomerie, and of course Tiger himself. Even LPGA player Natalie Gulbis is present, but the headline-grabbing Annika Sorenstam isn't, oddly.

Tiger Woods 2004 offers several features for new golfers. Based on the hole you're playing, an appropriate club is selected for you by default (you can choose a different one if you want.) "Caddy tips" provide hints about how to play each stroke of your approach. The commentary and suggestions from real-life broadcasters David Feherty and Gary McCord is engaging and professional, and really makes you feel like you're watching a Pro Tour on TV.

Instead of the usual 3-click swing interface used in many golf games, Tiger Woods 2004 uses a "real-time analog swing." To start your backswing, tilt the controller's joystick back; for the downswing, tilt forward. This introduces a small but interesting amount of skill into the game, and makes draws and fades off the tee fairly simple to do as an added benefit.

Past Tiger Woods games have had excellent graphics, and the 2004 edition carries on the tradition.

The character models for each of the golfers are highly detailed, and this year's Create-A-Golfer feature lets you precisely control how your character looks, which is lots of fun. Jaw size, head shape, eye color, even haircut and jewelry can all be tweaked to make a golfer who resembles yourself pretty closely. The animation uses high-quality motion-capture to perfectly mimic Tiger's unique playing style: his stance, swings, and signature victory moves.

Depending on the kind of mood you're in, you can tee off in a wide variety of modes. There's straight tournaments, challenge scenarios, stroke play for up to four players, head-to-head match play, and a basic practice mode. In the off-beat Skins mode, each hole has a dollar value, and you trade strokes for cash, forcing you to decide between maximizing winnings and topping the leaderboard. In Battle Golf, you can remove a club from your opponent's bag for each hole you win.

One nifty feature of Tiger Woods 2004 uses the clock inside your game console; when the real PGA Tour starts, you can play in your own virtual PGA Tour against the in-game pros. It's a great way to feel more involved with the actual Tour.

After you've completed a round, bask in the pleasure of your cash earnings, trophies, and medals. A host of scorecard stats are tracked for you automatically too, such as how many eagle threes you make on par fives. Finally, a virtual pro shop lets you spend your winnings on new clubs, balls, and other gear. *Tiger Woods PGA Tour 2004* is a deep golf game with a lot of replay value that any links fan would be happy to have.

Full-Swing Training Aids: Perfect Practice

Many golf teachers enjoy saying that "practice doesn't make perfect: practice makes *permanent*." By this they mean that working on correct fundamentals grooves a solid repeatable golf technique, whereas practicing the wrong ones (or the right ones wrongly) ingrains hard-to-fix swing flaws. Swing training aids, therefore, can serve as surrogate teachers, because they have the correct swing mechanics and technique designed right in to them and work kinesthetically to improve golfers' feel. People just have to go with the swing aids' flow, so to speak, and begin to register, record, and remember in their muscles' memory banks the motions and positions they teach.

Again, there are simply too many swing aids for this one chapter to list. Golfers will find many of those discussed in this chapter by searching Google or Yahoo! and entering the keywords "golf swing trainers." (For an excellent site that contains many trainers, go to www.golfaroundtheworld.com). The following roundup presents a sampling of the best swing training aids, both new and old. Of course, these devices aren't also called teaching aids for nothing. Many teachers use them extensively during their lessons (and many design their own), and, when portable and affordable, they encourage their students to purchase them and practice with them at home or on the range.

Another very true golf cliché states that a good swing starts with a good grip (see sidebar). Several molded or formed trainer grips on the market today use indentations and ridges to secure the golfer's hands in the correct position on the club. A training aid called the CoolGrip (www.coolgripgolf.com) has an electronic beeper built in to its plastic casing. The CoolGrip slips over a standard golf grip and emits a

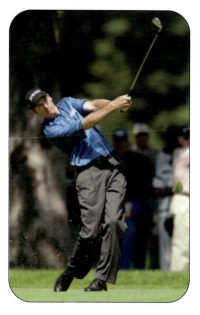

Jim Furyk at the U.S. Open

clicking bell-like tone when the golfer exerts too much hand pressure on the grip at any point during the swing.

Impact represents the most important position and alignment during the golf swing. A player may have a strange-looking motion, such as 2003 U.S. Open Champion Jim Furyk's (which TV commentators have described as "an octopus swinging in a phone booth," among other metaphors), but at impact the look and alignment of all accomplished players conform to a set of irrefutable golf laws, a flat left wrist and a bent right wrist (for right-handed players) and the clubshaft on the plane of the direction the golfer wants to start his or her shot representing the most important of them. That's how impact earned its moniker: the moment of truth.

A popular training aid is called the Impact Bag (www.golfaroundtheworld.com). It looks like a small ottoman and helps students ingrain proper impact geometry, both in the swing itself, and with equal importance, in their mind's eye. Students swing a club into a sturdy, although pliable bag, and then pose in the correct impact position. California-based LPGA teaching pro Debbie Steinbach calls it "the best training aid I ever found, because the bag gives the student instant feedback with regard to both correct body position and clubhead position at impact."

To further impress upon her students the importance of correct impact alignment, Steinbach takes a Polaroid picture of the student in this correct position against the bag, which allows students to study, see, and feel where their bodies and clubs need to be.

Again, a flat left wrist (for right-handed golfers) perfectly in line with the clubshaft and the left arm represents the first law of the correct impact alignment; the Impact Bag both teaches and reinforces this. Two other training aids focus on this wrist position *exclusively*. Golfers wear both the Tac-Tic Wrist trainer (www.tac-tic.com) and the Power Wrist trainer (www.powerwrist.com) on their left hands. The Tac-Tic clicks if the left wrist breaks down or cups (rather than remains flat) at impact, and the audible signal reminds golfers of their error. The Power Wrist trainer

The Tac-Tic Wrist trainer

features a firm piece of plastic that golfers wedge inside the backs of their golf gloves, and with which they can hit balls. The device, which does not allow the left wrist to collapse at impact, also has a pivoting hinge that allows the wrist and forearm to rotate properly through impact while maintaining the mandatory flat left wrist.

The golf swing, although circular in motion, actually takes place on a flat two-dimensional plane. (Actually, it usually shifts between two or more such planes.) Ever since Ben Hogan preached the importance of the swing plane, teachers and experienced golfers have searched for, discussed, and written passionately about ways of repeating an on-plane swing. Not surprisingly then, teachers and training aid makers have developed a plethora of devices and tools to help golfers swing on the correct plane. Well-known teacher David Leadbetter, for example, endorses the LazerGuide (www.laserguide.com), which projects a laser beam from a small device that golfers attach to the butt end of a golf club. Golfers see and know when they have made an on-plane swing when this beam perfectly traces a yellow graphic line painted on the LazerGuide's swing path mat on which the golfer stands.

Florida-based teaching pro Chuck Evans likes the Dual Track trainer (www.golfaroundtheworld.com), partly because his students can hit actual shots while using it. Evans points out that the golf swing has a backswing and a downswing plane, and this device trains students to shift between the two. To use it, golfers swing their club back along the device's two rails. When the club reaches waist high, it trips a lever that collapses the Dual Track, which falls to the ground behind the golfer. The rails allow golfers to make an on-plane backswing, then, after they fall away, golfers can swing back down and through the ball on a more inside path and a technically efficient shallower swing plane angle.

The Medicus 2000 (www.medicus2.com), which endorser Davis Love III first popularized years back, and the Whippy TempoMaster (www.tempomaster.com) remain two time-tested swing trainers that focus on the swing's rhythms, and many teachers continue to use them. The Medicus 2000 club has a two-way hinge in the middle of its shaft, which collapses the club into halves when swung with a jerky and uneven motion. The Whippy TempoMaster club teaches rhythm, timing, and tempo throughout the swing. It has a thin and extremely flexible wiry shaft that golfers must swing smoothly; otherwise, the club will fly immediately out of its uniformed orbit and swing path, and sometimes even hit them in the head.

David Leadbetter, founder of the David Leadbetter Golf Academy, on golf teaching aids

If video systems help golfers visually, training aids help them by focusing their attention on feel. In other words, what we're trying to do with a teaching aid is simply improve golfers' kinesthetic awareness.

Normally, the best training aids are the simplest ones, because most people are a little uncomfortable going out on the practice tee with something strange looking. "The Leadbetter Glove," which is one of the first training aids I designed, is very simple in its idea and is still a very popular aid, because everybody has a problem with their grip. The product has very clear black lines drawn on it that correctly place the club in the student's left hand (for right-handed players). It's approved of by the USGA, so people can wear it while playing on the course or in a tournament.

Simply put, a quality teaching aid can help students create good habits. It can give them a better physical awareness or "feel" for when they are doing something wrong or doing right in the swings.

Short Game Training Aids: Bogey Busters

Golf pundits like to call putting a game within a game. Instead of hitting the ball a mile through the air toward a big fairway or green, golfers roll the ball into a 4.25-inch diameter hole. Instead of generating power through a full-body pivot, good putters anchor their bodies motionlessly and quietly stroke the ball using only their shoulders, arms, and hands. The chipping and pitching stroke lies along a very real continuum between the putting stroke and the full golf swing. Remember, statistics show that golfers play between 65% and 70% of all their shots during a round from 130 yards or closer to the pin, so all would do well to improve their short game skills.

Like putters and putting strokes, putting training aids display a sometimes quirky range of imagination in an often eye-popping array of shapes and styles. In fact, in short game guru Dave Pelz's book, *Dave Pelz's 10 Minutes a Day to Better Putting*, Pelz guides readers through scores of putting drills and exercises, many of them using putting training aids that he developed. In a sidebar in this chapter, Pelz discusses his philosophy about putting training aids and also describes a few of his newest creations.

First, however, here are a couple of other putting trainers that have withstood the test of time and continue to help golfers improve their scores.

For years teachers and golf students have worked with a simple putting trainer called the Putting Connection (www.golfaroundtheworld.com), which also works as their chipping aid. Comprised of a simple plastic adjustable bar that fits between the golfer's arms, the Putting Connection locks the elbows into a "V" position at address and keeps them there throughout the putting or chipping strokes. This added structure helps students develop a smooth-pendulum stroking action, where they control the putter by moving their shoulders and arms, and not by flipping their hands and wrists.

A new trainer called the Putting Arc (www.theputtingarc.com), offers a variation on the classic putting track, which guides the putter on a straight-back/straight-through path. Handsomely crafted from stained wood, this product teaches a putting swing path that moves the club slightly to the inside of the intended putting line on the backstroke and then back again inside on the line on the follow through. In other words, the golf pros who helped develop the aid think the best putting path inscribes an arc, not a straight line. This makes sense because the center of the putter, called the "sweet spot," and the putter's shaft actually lie on two separate planes. Swinging the shaft of the putter straight back and straight through actually pushes the sweet spot plane outside of the shaft's plane, whereas swinging the putter along the putting arc's edge as

intended moves the sweet spot almost straight back and through the ball. What's more, the manufacturers say they developed this product by using a mathematically verified formula developed more than 2,000 years ago by Apollonius of Perga, the Greek scholar who was known as the "Great Geometer" (but could he putt?).

As the old song says, there's nothing like the real thing, except, maybe, *something* like the real thing. Golfers obsessed with improving their putting can do so on one of the several quality artificial putting greens now on the market, such as the SofTrak personal putting green from United Turf Industries. Made of special soft but dense synthetic fibers, the green has the look, feel, and performance characteristics of a green built to USGA standards. It can be customized for size, speed, and contours and can receive chip, pitch, and even full approach shots with irons and woods. The average cost of a 400- to 600-square foot green ranges from $7,500 to $12,000, depending on its features.

The Putting Arc

The SofTrak artificial green

Some of the same training products described in the full-swing trainer sections work equally well to improve golfers' chipping and pitching skills. These shots often fail because golfers anxiously scoop at the ball with a cupped or broken down wrist, rather than a correct flat left wrist, at impact. Golfers can work to eliminate these flaws with trainers such as the Tac-Tic Wrist trainer and the Power Wrist trainer, and the Impact Bag.

Dave Pelz, acclaimed short game teacher, training aid inventor, author, and Director of Pelz Golf golf schools and clinics (www.pelzgolf.com)

Dave Pelz is acknowledged worldwide as a leading authority on the short game and putting. He studied physics at Indiana University before spending 14 years as a space research scientist at NASA's Goddard Space Flight Center. He then came back down to earth and turned his attention to helping people get a little white ball into a small black hole.

Pelz says that learning to be a great putter is all about learning to interpret accurate feedback. "If you ever put your hand on a hot stove, you probably never did it again," he begins, "because the experience provided very accurate and reliable feedback." Pelz goes on to say, "When most golfers miss a putt, they can't interpret the feedback the missed putt gave them, so they have no idea why they missed it. That means they have no idea about how to get better."

Pelz points out that physically, mentally, emotionally, and in every other way, putting a golf ball is simpler than hitting a full shot. "But," he adds, "putting is harder to learn." He says that's because people think that if they putt the ball and it goes in the hole they did something good, and if the ball doesn't go in, they did something bad. Certainly a missed putt represents feedback, Pelz acknowledges, "but it doesn't represent very accurate feedback."

He explains that one of the problems all golfers have, pro and amateur alike, is that they tend not to read the entire putt. Instead, they usually only read the last part of their putt, where the ball makes its final break toward the hole. According to Pelz, golfers basically ignore all the feedback most of the entire length of the putt could have given them.

Pelz, assisted by Tour star Phil Mickelson, tried to remedy this fault by developing a putting-alignment/green-reading feedback device called the Putting Tutor.

"It's a triangular plate that has a straight line running through it and that you lay on the green," Pelz begins. "The device also has two brass marbles, which form a gate just 8 inches in front of the ball and which the ball will barely fit through. You aim this line where you think you're going to start your putt and only if you start that ball on the line of your aim will it roll cleanly through the marble gate without hitting the marbles. So you get clear and accurate feedback about your ability to aim the entire putt right from its starting point."

Pelz has developed all of his putting aids with the philosophy that he doesn't want them to interfere with the mechanics and feel of a well-executed stroke. If a golfer makes a mistake in the stroke, however, the trainer will let the player feel that mistake. After all, Pelz points out, golfers play with feel on the golf course, not mechanical appendages, which, of course, the USGA deems illegal during competitive play and in rounds used to establish a handicap.

Actually, Pelz's ingenious new putting aid called the O-Ball doesn't violate golf's rules, so golfers can actually compete with it. This top-quality multiplayer ball has two parallel red O rings painted on it. After aiming the rings at the target, the golfer strokes his or her putt. If golfers stroke it exactly where they have aimed, the ball will roll without the red lines wobbling at all. If they cut across the ball or hit it on the toe of the putter, or do something else wrong, however, the lines on the ball wobble like mad, which gives golfers instant feedback that they didn't make a very good stroke or hit the ball solidly.

The O-Ball

The LazrAimer

Finally, Pelz discusses a training aid with a more little high tech in it. "I designed the LazrAimer trainer to be the size of a golf hole that you put on the floor," says Pelz. "You attach a tiny little Plexiglas mirror on the putter face and you aim the putter at the LazrAimer as if you were aiming at an actual hole. When you think you are aimed perfectly, you say the word 'on' and a voice-activated circuit in the LazrAimer turns the laser on. A beam comes out of it and hits the mirror on the putter and then bounces back again. If the return beam doesn't hit the LazrAimer when it bounces back, the golfer has not aimed correctly. It's as simple as that."

Pelz concludes by making the point that although all great putters have excellent mechanics, they aren't great putters until they have developed feel and touch, which they do through practice. "Tiger Woods works on his mechanics all the time," Pelz says, "but when he's playing the final hole of a tournament, he's not thinking about mechanics, because he's practiced them enough to make them automatic and subconsciously controlled. If a player can't get past mechanics, he or she will never be the best player he or she can be." In concluding, Pelz reminds golfers that they develop feel and touch not only on the putting green, but also by putting out on the golf course under actual playing conditions.

The Healthy Golfer: Physical Fitness on the Links

Tiger Woods, who swung a golf club before he could even walk, has done more than inspire his succeeding generation of talented kids to take up the game of golf. Woods's phenomenal athleticism has also confirmed golf as a legitimate sport. That said, although Hall of Famer Gary Player has lifted weights and trained rigorously for golf for more than 40 years, the fitness craze has only recently crossed over into the world of golf. Today several Tour players, even the young "flat-bellies," travel with personal trainers and work out regularly at tournament sites in Tour-sanctioned fitness vans that travel the circuit. Physical therapist Rob Mottram worked in the PGA vans for 11 years. Now he develops golf-specific fitness programs for both aspiring professionals and recreational golfers at his Golf Health and Performance Center in Palm Desert, California (see sidebar).

Many swing trainers also double as golf-specific strengthening devices. The Swing Wave, for example (www.swingwavegolf.com), works on improving both a player's strength and his or her storage of power during the swing. To use it, golfers fill a tube-shaped container with water, which attaches to the trainer's shaft. On the backswing, the water falls to the bottom of the container. Golfers make sure they keep it there both starting down and well into their downswing. When they reach the hitting zone, they release the club as the water swooshes forward through the container.

Tour star David Duval popularized the Momentus trainer (www.momentusgolf.com), a heavily counterweighted club, with weights on the handle and in the middle of the shaft, that not only builds and tones muscles, but also helps golfers swing on plane. Top teachers understand that for golfers to hit the ball farther, they must not only swing on plane, they must increase their swing speeds. To help them do this, noted teacher Jim McLean has

Jim McLean

The McLean Power System

brought out his McLean Power System (www.swingrite2.com), which consists of three weighted clubs designed to increase power and swing speed through a series of strengthening exercises.

A very interesting organization called Back to Golf endorses the Coach (www.backtogolf), a gym-quality exercise apparatus that David Leadbetter also helped to originally develop. It builds and stretches golf muscles and assists in grooving a technically sound swing. Physical therapist Bud Ferrante and PGA teaching pro Tom Nix started Back to Golf in

1991. The Fresno, California, based company, among other services, conducts seminars across the country each year, educating physical therapists and golf teaching professionals about the biomechanics of the golf swing and how to design and implement rehabilitation and performance programs specifically for golfers.

North Carolina-based LPGA teaching pro Karen Palacios-Jansen combines aerobics, strengthening, and swing training in her Cardiogolf videos and classes (see sidebar).

Tour players and amateurs alike know from experience that physical conditioning can make the difference between driving the ball down the middle of the fairway on the 18th hole or slicing it weakly into the woods. That their commitment to golf conditioning reduces stress, gives them more energy, and allows them better focus *away* from the golf course stands as just another birdie on the scorecard of life.

Rob Mottram, golf trainer and physical therapist, owner of Golf Health and Performance Center (Palm Desert, CA), on golf fitness and technology

Fitness and golf really didn't coexist (with the exception of a few players like Gary Player and Frank Stanahan) until the mid-1980s. That's when Dr. Frank Jobe, the world-renowned orthopedist, cofounder of the Fitness Institute and Medical Director of the Biomechanics Laboratory at Centinnella Hospital in Inglewood, California, began his biomechanical research and analysis of the golf swing. Again, until then hardly any Tour players worked out with weights, because they thought it would make them muscle-bound or overly tight (and this would restrict and therefore hurt their swings). All that has changed, in part, because we now have scientific facts that support the effectiveness of golf-specific training.

The first thing we do at Golf Health and Performance is figure out what a golfer's body specifically needs to become stronger and more healthy. Every golfer is different in terms of flexibility, strength, balance, coordination, speed, and old injuries he or she may have sustained. We address these matters first because they influence and determine the type of training program we prescribe for that person. When golfers begin working with us, and they improve their balance and develop a good fitness foundation, we go into things such as the overall strength and speed components of the swing and do some specific swing-skill work.

Technology greatly helps us achieve our goals with our golfer clients. For example, we use a wonderful motion system computer that a Phoenix-based company called Skills Technologies developed. It works by attaching four electromag-

A golfer wired for a high-tech physical fitness evaluation

netic sensors to the body of the golfer, which creates a 3D animated virtual model of that player swinging a club. The computer tells us how fast the sensors are moving through space, and from that we calculate several things about the golfer's lower body, upper body, and arms, and the golf club. The computer tells us where each one is in space through the swing, how fast they are moving, and in what order they fire or move. We've measured enough world-class players so that we know what the pattern should be.

The golf swing takes only two seconds to complete. If you're on an older exercise machine that has a tempo of 3/3, or three counts out and three counts back down, however, you haven't trained the nervous system to turn on the muscles at the high rate of speed that you want for golf. So we've gotten into plyometrics, which is an explosive advanced method of training that Eastern European Olympic athletes have used for many years. The idea here is that you load the muscle and it springs back. A good example is a person standing on a chair and then jumping to the ground, then immediately jumping back up again. As that person jumps to the ground, he or she loads the muscle when hitting the ground and before springing back up. So the muscle has a rebounding effect. Another example is throwing a heavy medicine ball to a golfer and having him or her throw it right back. In golf, when you take the club back, golfers stretch all of their muscles. Then, before the golf club comes to rest at the top of the swing, golfers push off with the lower halves of their bodies, and that's where they get that load-and-rebound effect through the downswing that produces power.

Karen Palacios-Jansen, North Carolina-based LPGA teaching pro, on her Cardiogolf fitness and swing training program. (www.swingbladegolf.com)

Karen Palacios-Jansen

I used to have a dilemma: I could spend my free time either working out or working on my golf swing. Golf may be good mental exercise, but as far as physical exertion is concerned, it can't compete with aerobics. I got to thinking, maybe I could do both at the same time, so I developed an exercise program called Cardiogolf. Cardiogolf is a golf-specific exercise program I designed to help players improve their swings by conditioning their muscles.

This program is designed for golfers of all levels, regardless of age or level of play. Cardiogolf teaches people to condition the specific muscles used in proper golf swing biomechanics, and provides low-impact aerobic conditioning, too. Cardiogolf is built around a 45-minute workout that combines drills and exercises for improving swing technique and for increasing clubhead speed, which is a prime ingredient for hitting the golf ball a long way. By improving swing technique, golfers not only will perform better on the golf course, they will also decrease their chances of golf-related injuries. People often overlook and underestimate the importance of aerobic conditioning for golf; today, however, almost every Tour player works on improving his or her endurance through aerobics.

After a warm-up of stretching, Cardiogolf moves into an extended cardiovascular routine that uses the "Perfect Cardio Club" and really gets the heart beating. The Perfect Cardio Club is a shorted club designed specifically for this indoor workout. To begin the continuous-motion exercise program, we start by working on the golfer's setup, posture, and ball position, and then we progress into the weight shift and maintaining the spine angle through the swing. Next we add the right arm, the left arm, both arms, and then move into exercises for increasing swing speed. The workout also uses hand weights for strength conditioning, and closes with a cool-down routine.

Fun in the Sun: Sunglasses and Sun Protection

A couple of generations ago, when Curtis Strange, Greg Norman, Nick Price, and Peter Jacobsen turned pro, virtually no one wore sunglasses on Tour. In fact, much like the stigma attached to heavy exercise training that kept golfers out of the weight room until the early 1990s, golf pros thought tinted lenses would distort their perception of the golf course and hurt their games. Few seemed concerned about damaging their eyes by spending each of their working days in full sunlight. In fact, it took a quality-sunglasses maker like Oakley (see sidebar), whose products were already popular with athletes in many other outdoor sports, to crack the golfer barrier. When Tour stars David Duval and Annika Sorenstam donned shades about 10 years ago, other Tour players quickly followed. Today, vastly improved lenses and frames from Oakley, Bolle, Nike, adidas, Maui Jim, and other manufacturers not only protect golfers' eyes, they both enhance their perception of the golf course and their appearance for their adoring fans.

Golf-specific sunglasses from these companies feature a host of technologies that promise minimal distortion, accurate depth perception, startling clarity, and, of course, 100% protection from harmful UVA, UVB, ultraviolet, and blue-light rays. Glare-reducing tints available in interchangeable lenses let golfers match their sunglasses with the amount and quality of sunshine on any given day, while producing ideal color spectrums for the game of golf. Violet tints, for example, highlight and enhance the color green, which allows golfers to see the grain in the greens and the undulations in the greens and fairways more acutely and accurately than with the naked eye.

Featherweight, thin, high-tech plastic and titanium frames that barely kiss the golfer's skin create a pressureless and comfortable fit. Rimless bottoms

eliminate visual obstruction and come very close to making golfers forget they even have sunglasses on. What's more, these frames hold their shapes as well as resist sun and salt corrosion and last for years.

Scratch-resistant polycarbonate lenses resist the impact of sand, small stones, or other debris kicked up into them by a club striking the ground at impact. So, when staunch traditionalists such as Nick Faldo and Justin Leonard show up on the links with golf shades on, average golfers have no excuse whatsoever not to wear them when they play.

Of course, golfers must protect their skin as well as their eyes from the dangers of the sun. This is particularly important for young golfers, because much of the damage that occurs from excessive sun exposure happens in childhood, although the dangerous effects (skin cancer, premature aging) often do not present themselves until adulthood. A company called Proderma (www.prodermaproducts.com) has developed a line of nongreasy, fast-absorbing sun-protection lotions, creams, and waxes especially formulated by an avid golfing chemist with golfers in

The Proderma products

mind. These products do not interfere with a secure grip on the club and come neatly contained in a compact case that fits into a golf bag. They include basic SPF 30 sun- and windscreens (including a separate full-body sunscreen), SPF facial moisturizer and body lotion, SPF face stick for areas around the eyes, and also a Moonscreen, which helps to bind damaging free radials and inhibits their release into the skin.

Proderma has recently launched their "ForeWard" educational campaign intended to raise awareness about and reduce the incidence of skin cancer contracted by golfers in the fall and winter.

Louis Wellen, Director of Sports Marketing, Oakley, on the technology of Oakley's golf sunglasses (www.oakley.com)

Sunglasses from Oakley

Back in the 1980s, the industry said that you couldn't take a polycarbonate lens, which is a high-tech plastic, and make it so that it is optically clear to see through. Not only did we prove them wrong, we also make polycarbonate lenses that are nearly optically perfect to look through. Prior to that, the experiments in polycarbonate lenses didn't make the geometric corrections in how the light passed through the lenses so that they would provide optically correct vision. Before we came along, everyone was making flat glass lenses, so that when light passed through them, there was no distortion. Now if you take something and put a curve on it and pass a beam of light through it, it will curve and distort it. We developed what we call XYZ Optics by molding the lens in the shape of the human eyeball and making the lenses taper-corrected. This means that the lens is thicker closer to the nose and gets thinner as it moves toward its edges, which prevents the distortion of light passing through. In fact, this breakthrough design began with our first sports sunglasses, called the Blade, a wraparound lens (which evolved into the Shield) that was revolutionary at the time. That cylindrical lens shape curved from left to right. Today we design our lenses off of a spherical lens blank, which means that it curves from left to right and from top to bottom, and, again, this mirrors the shape of the human eyeball.

Now let me say that American society is probably about 20 years behind the rest of the world as far as eye and skin protection from the sun is concerned. In Europe, New Zealand, Australia, and Japan, for example, there are public service commercials on TV every 30 minutes telling people to protect their eyes and skin. An Australian ophthalmologists organization sponsors such a campaign, whose commercials say "slip, slop, slap," which means slip on a pair of sun glasses, slop on some sunscreen, and slap on a hat.

When we first started encouraging professional golfers to use our eyewear, I predicted that within 5 to 10 years, we would have 50% to 60% of the Tour playing in eye protection. It's been about 10 years, and we haven't quite reached 50% yet. But I'll tell you what, every time I'm out at a professional tournament, the players who want to try playing in sunglasses bombard us with requests for our products, because their eye doctors are telling them they need eye protection.

Golf Apparel: Clothes Maketh the Golfer

Although a golf course has clear boundaries that separate it from the surrounding world, golfers want to cross those borders in clothes that work well for them on both of its sides. Indeed, as the game's popularity has grown and spread through different age groups, occupations, economic brackets, and geographical regions, golf apparel has expanded its design horizons to appeal to a broader range of people's tastes and styles as well. What's more, golf's growing status as an athletic sport requiring strength and speed has led to technological innovations in golf apparel designed to augment the golfer's freedom of motion during the swing.

Like heroic mailmen and women, some golfers will let nothing deter them from completing their appointed rounds. Thankfully, advances in apparel and outerwear continually challenged the forces of nature, and, like the sensitive climate-control systems in expensive cars, keep today's golfers comfortable in spite of the weather. Rain suits even come with waterproof zippers, which allow outerwear makers such as Sunderland of Scotland (and many other fine ones (including Polo, Titleist, and Zero Restriction) to literally assert that their customers can play 18 holes in the rain and stay completely dry (see Paul Sunderland's sidebar).

It should surprise no one, then, that athletically minded companies such as Nike and adidas, who, for years, have manufactured high-tech apparel for tennis, skiing, running, and other sports, have led the way in technology/performance-based golf clothes, with other companies quickly following suit.

The Golf Shirt: A Classic Redefined

Technology has had a huge impact on contemporary golf clothes, both in the men's and women's market, starting with the new synthetic technical fabrics companies use in making their clothes. These include lightweight blends that stretch, breathe, and are water resistant and sun protective, such as the sporty Nike mock turtleneck Tiger Woods often wears on the PGA Tour. Call it dri-Fit from Nike, ClimaCool from adidas (see Paul Clark's sidebar), PlayDry from the Greg Norman Collection (among other offerings), each company has its own moisture-wicking nomenclature, but all take on and defeat heat and humidity and share a commitment to style and performance.

Nike Golf's Dri-Fit mock turtleneck

Companies continue to make progress integrating UV protection into their garments, and in making them porous enough to let moisture and heat move from the body outward, while sufficiently opaque or dense to block harmful sunrays from getting in. In addition to Tiger's mock turtleneck, other manufacturers now use anti-UV technology in their long-sleeve turtlenecks, windshirts, and pullovers intended for cool weather when breatheability is less of an issue.

No one today will mistake a golf shirt for a tent. As David Hagler, Director of Apparel, Nike Golf, points out, in years past golf-shirt makers cut oversized garments to facilitate the golf swing's movement. Today they integrate Lycra and other fabrics into shirts, which stretch with the golfer's body during the swing and offer a more athletic fit and look. Manufacturers have moved shoulder seams on shirts lower toward the chest, which both keeps them from interfering with the swing and from irritating the shoulders of golfers carrying their bags. Stretch fabrics have also found their way into golf pants and shorts, and, again, this helps golfers move or pivot more comfortably through the swing.

Hagler adds that women have worn better-fitting athletic clothing that tapers more snugly to their bodies for some time. As for the future of golf apparel, Hagler thinks that in a few years golfers might look for "smart" phase-changing fabrics (already used in gloves and hats), that, like a personal thermostat, cool them down in hot weather and warm them up when it turns cold.

Tiger Woods wearing Nike,
Copyright Getty Images

Paul Clark, General Manager, adidas Golf, on golf apparel and technology

Building a golf shirt used to be a fairly simple project because if you found a source experienced in making golf polos, you could get in the golf-shirt business. What sets adidas and a few other companies apart in today's world of golf apparel manufacturing, however, is the technical benefits our products display, which have been revolutionary in the past few years.

Hidemichi Tanaka for adidas

A discussion of contemporary golf apparel might begin with a company such as Ashworth. They may have been what you would call the first authentic golf apparel brand, with their 100% cotton piques and oversized cotton shirts, which allowed freedom of movement in the golf swing. Now the trend has moved away from cotton altogether, which is really quite a move. Let me put it this way: Just three years ago the golf apparel market was 100% cotton; and if you said then that in three years the cotton polo golf shirt would be on its way out in favor of synthetic blends, some people would have laughed at you. But this is exactly what has happened.

At adidas, for example, we have golf shirts that blend cotton and polyester microfibers, with the synthetic fabric on the inside that touches the body and cotton on the outside. The synthetics absorb and wick, or pull the moisture away from the body. It then escapes through the cotton weave before evaporating into the atmosphere, allowing the golfer to stay cool and very dry. Our new ClimaCool line is made from 100% synthetic fabric, for the ultimate in lightweight, comfort, and breatheability. One of the knocks against the synthetic fabrics has been that they retain odor, but we've worked with some vendors now to develop 100% synthetic fabrics that breathe extremely well and wick the moisture away but do not retain body odor. Not only that, but also these shirts weigh about half that of an all-cotton polo. So on a 95-degree day while playing golf in Arizona, golfers stay completely dry and have such freedom of movement they feel as if they are wearing nothing at all. It's a pretty interesting concept that we've realized in our products.

Outerwear and Rainwear: Neither Rain, Wind, nor Snow...

Nobody can predict the weather or the outcome of a round of golf. On some days, the swing feels smooth as if the sun will never stop shining. On others, a hard wind and a driving rain turn a game meant for fun into a long, hard march. Little, however, keeps the most steadfast golfers from wandering off course, especially when decked out in one of today's vastly improved rain suits, wind jackets, or fleece-lined pullovers. These garments feature incredibly lightweight and quiet fabrics, laminated with water repellent or waterproof coatings (in some garments combined with wicking liners) that keep golfers dry, cool, and comfortable without restricting their swings. Indeed, previous generations of outerwear often made golfers feel like they were swinging their clubs in a steam room jumping with noisy and screechy crickets.

PGA Tour player Billy Mayfair speaks for every golfer wearing such gear when he describes his new rain suit as "a hundred thousand times better than the one I played with 10 years ago." He points out that Davis Love III wore a rain jacket during the final round of the 2003 Player's Championship

Tournament, which Love won on a hot, humid, and intermittently rainy Florida day, even when it wasn't raining. Love won the tournament and during a televised interview afterward said he just about forgot that he had the jacket on.

Most of the major golf apparel manufacturers offer quality outerwear and rainwear. Price ranges vary, with tags such as "water-resistant" or "repellent" signaling considerably less-expensive and element-protecting garments than a "waterproof" designation. (Waterproof garments are by definition also always windproof.) As in golf shirts, many pieces of rainwear feature moisture-wicking fabrics that mitigate perspiration.

Paul Sunderland, President of Sunderland of Scotland, knows a thing or two about rainwear. His great-grandfather founded a tailoring company in Scotland in 1911, which entered the golf apparel business in the 1920s. Paul's father, also known as "the father of golf outerwear," began manufacturing rainwear at Sunderland for golfers in the late 1950s. Today the company exclusively provides outerwear and rainwear for both the European and American Solheim Cup teams, and continues to make fine products at a range of price points today (see Paul Sunderland's sidebar).

Paul Sunderland, President of Sunderland of Scotland, on golf rainwear

The waterproof Viper fiber, which we use on our GT Tour rain jacket, is a breakthrough material made by Gore-Tex. It represents a major innovation in rainwear, because it both stretches and is completely waterproof. We sew this material into the shoulder, elbow, and sides of this garment, and because it has the stretch-enhancing Lycra in it, golfers wearing it experience 100% freedom of motion. A Gore-Tex laminate makes up the rest of the jacket, so golfers stay completely dry and cool. In years past, manufacturers had to put a lot of effort into attaching elastic sections on various parts of their rainwear to increase its flexibility. This made the products

A Sunderland jacket

heavier, bulkier, and restricted the golfer's swing in awkward ways. Viper and fabrics like it eliminate this problem, because they stretch and are completely waterproof at the same time. Golfers have to pay for our products with Viper, but those who want the Rolls Royce of rainwear happily do so. At a bit of a lower price point, we see microfiber outer shells with waterproof inner linings.

With the next breakthrough in rainwear, golfers will feel like they are not wearing rainwear at all. In fact, today's rain garments are so well tailored that people now wear them about town on sunny days.

The Mental Game: A Psychological Tool Shed

Legend has it that during an interview a golf writer asked a seasoned Tour star what percentage of the game of golf he thought was mental. The player immediately responded, "90%." Then he just as quickly reconsidered his answer and revised it to 100%. Although that doesn't leave much room for the clubs, the ball, the course, or even the golf swing, it does indicate just how much world-class players value the psychological side of golf. In fact, both Jack Nicklaus and Tiger Woods have publicly rated their mental toughness and golf course management skills as the strongest parts of their games.

During the past decade, many Tour players added sports psychologists to their team of teachers and consultants. Some, like the well-known Bob Rotella, have achieved a celebrity status all their own and share their insights into shoring up the mental side of the game in top-selling books such as *Golf Is Not a Game of Perfect.* Another is Pia Nilsson, a former LPGA Tour pro, who, when she directed Sweden's competitive golf program, helped Annika Sorenstam develop her all-around game. Nilsson, like her business partner, top-rated teaching professional Lynn Marriott, founded and directs an exciting and innovative teaching/learning/publishing organization called Coaching for the Future. With the help of cutting-edge technology, they work with their players to help integrate the physical and mental aspects of golf (see Pia Nilsson's and Lynn Marriott's sidebar).

LPGA teaching professional Maxann Shwartz holds a Ph.D. in clinical psychology. Dr. Shwartz teaches golf at Strawberry Farms Golf Club in Irvine, California, and is the psychologist affiliated with Debbie Steinbach's multi-dimensional golf organization Venus Golf (www.venusgolf.com), where she is known as "Dr. Max." Dr. Shwartz administers psychological assessments to golfers of all ages and children in educational settings to maximize performance. She also uses her knowledge of standard psychological tests to analyze her golf clients' basic psychological profiles to help diagnose disorders such as Attention Deficit Hyperactivity Disorder (ADHD) and others that may be adversely affecting their golf games (see Dr. Maxann Shwartz's sidebar).

Dr. Deborah Graham working with a client

Golfers wanting to fine-tune their golf cognitive skills online might connect with sports psychologist Dr. Deborah Graham's Golfpsych web site (www.golfpsych.com). Also a Ph.D. in clinical psychology, Dr. Graham and her husband, John Stabler, have compiled their research into a fascinating book called *The 8 Traits of Champion Golfers*. People get a rudimentary idea of where they fall in a performance continuum of world-class players by taking a sample test for free online. The test offers a kind of golf-specific psychological profile that gives people an idea of what cognitive and emotional skills they need to improve to play their best. The company charges for a much more in-depth version of this test, which includes a response in the form of a 50-page personalized report. Golfpsych has developed their own portable Tension Meter, which monitors the heart to calculate a golfer's arousal level and capacity to achieve ideal psychological states for optimal performance.

Dr. Maxann Shwartz, on psychological testing of golfers

As a psychologist, I use technology to help gain insight into personal strengths and weaknesses in psychological functioning, whether it is to help a seasoned Tour professional, a new golfer, or a child having difficulty learning to read. Through the use of psychological and neuropsychological assessment tools, it is possible to identify how the individual processes information most efficiently and where specific deficits may or may not exist. From a psychological perspective, a thorough assessment can be a valuable tool that enables a person to gain more understanding in how his or her body-mind functions most optimally.

For example, several computerized tests can aide in the assessment of attention and/or impulse-control problems. This information is used in the context of a number of other assessment methods (for instance, historical information, current performance levels, rating scales, observations, and self-reports). Through the use of psychological technology, weaknesses and deficits can be identified, too, which may help the individual find positive interventions.

After several psychological tests and interviews, one particular 10-year Tour veteran discovered that she had severe Attention Deficit Hyperactivity Disorder (ADHD). This didn't come as a surprise to her and, in fact, made sense as to why historically when she was in a position to win a golf tournament, she mysteriously lost focus and was unable to filter out unnecessary stimuli (leader board, crowds, and so on). She always felt as though she "choked" when in reality her brain chemistry worked against her in highly stressful competitive situations. Although normally this phenomenon is termed "competitive anxiety" and is experienced by everyone who has been in the position to win, it cannot be argued that some athletes handle this level of stimulation better than others. This is a perfect example of how brain chemistry happened to work against this particular golf professional. Had she learned of her condition earlier, she may have undergone treatment. ADHD is a viable medical disorder; however, it often receives bad press in that it is often overdiagnosed. In this golfer's case, it was undiagnosed and cost her dearly.

I also have used psychological testing (which really refers to the "technological side of psychology") to help a seasoned LPGA Tour veteran who had experienced several difficult years on Tour. A past Tour winner and holder of several impressive records, this golfer was assessed with a full battery of tests. Results of the testing revealed that her ability to use visual processing pathways was much weaker than her ability to use auditory or verbal modes of information processing. Consequently, when this was explained to her based on her testing scores and patterns, she indicated that she had always had difficulty visualizing her target and, in fact, had worked with several sports psychologists but just could not visualize. Because her auditory processing skills were clearly stronger and, in fact, her scores fell in the high average to superior range, together we devised verbal cues designed to help her with her shot making. The golfer came up with the words and phrases (which were very short and simple) and began to rely on more verbally loaded modes of self-communication instead of visual techniques (that is, imagery or visualization). After practicing this technique, she won the very next professional event that she entered.

This is one clear example of how a very experienced golfer was able to benefit from advances in technology! As a psychologist, I believe that advances in the technical side of psychology can and will have a positive impact on performance in not only golf, but also in all sports. By better understanding psychological functioning and its effects on performance and skill acquisition, how can the golfer not benefit? At worst, the individual may learn something new about himself or herself.

Pia Nilsson and Lynn Marriott of Coaching for the Future, on technology and the mental game (www.coachingforthefuture.com)

It is our intention to use technology to support the experience of the players we work with, so they can transfer an awareness of their games and themselves to the golf course. We want all golfers to become more aware of their swings, mental states, and emotional states so that they can experience and adjust to them as needed during the round. This awareness has to be something the player experiences internally. We have found much of the technology (specifically video swing analysis systems) today creates an unbalanced dependency within players, because they become too dependent on something external that shows them how they are "supposed to do it." In other words, too many players have become dependent on the video screen and its graphics, without a balance of internal awareness.

We have found swing changes that can "hold up under pressure" have to be experienced implicitly by the players themselves and become a part of their preshot routine, without disrupting their ability and capacity to connect to and focus on the target.

Nevertheless, we have found a video technology that we think is very good and helpful. Our Coaching for the Future's GOLF54 program uses the Swing Solutions GVA600 video system that has been integrated with a company called HeartMath and their Freeze Framer Interactive

Pia Nilsson and Lynn Marriott

Learning System. The Freeze Framer technology is based on more than a decade of research on the relationship between the heart, health, and performance. Players can watch in real time how their thoughts and emotions affect their heart rhythms; and by seeing the changes in their heart rhythms on the screen, players can learn to quickly balance their emotions, mind, and body. Ultimately they can learn how to access "the zone" and train themselves to stay in the zone for optimal health and performance.

By using the latest technology available, our goal is to provide an environment whereby players can experience the integration of the physical, mental, and emotional sides of themselves and become their own best coaches. Remember, peak performance can be sustained only from an awareness cultivated from the inside out.

GOLF54 represents our belief that we as human beings can score a lot lower than we do today. GOLF54 is about removing the barriers we set for ourselves and moving toward a score of 54 during a round of golf and beyond. What do we need to know? What do we need to learn? What is our vision? How do we need to practice? How do we need to think and act? What do we need to believe in? We believe that all our potential is enormous and unlimited.

Resources

Section 1: The Golf Gear

Golf Clubs

Adams Golf
2801 East Plano Parkway
Plano, TX 75074
(www.adamsgolf.com)

bellyputter.com
(www.bellyputter.com)

Burrows Golf
28144 Harrison Parkway
Valencia, CA 91355
(www.burrowsgolf.com)

Callaway Golf
2180 Rutherford Road
Carlsbad, CA 92008
(www.callawaygolf.com)

Cleveland Golf (Never Compromise Putters)
5630 Cerritos Avenue.
Cypress, CA 90630
(www.clevelandgolf.com)

Cobra Golf
P.O. Box 965
Fairhaven, MA 02719
(www.cobragolf.com)

Frankly Golf
31 Fairmount Avenue
PO BOX 707
Chester, NJ 07930
(www.franklygolf.com)

Kit Mungo: A Better Club (club repairs)
6650 Crescent St. #1
Ventura, CA 93003
(www.abetterclub.com)

La Jolla Club (LJC) Golf
2445 Cades Way
Vista, CA 92083
(www.ljcgolf.com)

Lange Golf
15866 W. 7th Ave., Suite E
Golden, CO 80401
(www.langegolf.com)

MacGregor Golf
1000 Pecan Grove Dr.
Albany, GA 31701
(www.macgregorgolf.com)

Mizuno
4925 Avalon Ridge Parkway
Norcross, GA 30071
(www.mizunousa.com)

Nancy Lopez Golf
18 Gloria Lane
Fairfield, NJ 07004
(www.nancylopezgolf.com)

Nike Golf
One Bowerman Drive *503 671 6453*
Beaverton, OR 97005
(www.nike.com/nikegolf)

Odyssey Golf
2180 Rutherford Road
Carlsbad, CA 92008
(www.odysseygolf.com)

Orlimar Golf
1385 Park Center
Vista, CA 92083
(www.orlimar.com)

PING
P.O. Box 82000
Phoenix, AZ 85071
(www.pinggolf.com)

Scotty Cameron, The Art of Putting
(www.scottycameron.com)

Sonartec
2720 Loker Ave West, Suite-A
Carlsbad, CA 92008
(www.sonartec.com)

Square Two Golf, Women's Golf Unlimited
18 Gloria Lane
Fairfield, NJ 07004
(www.squaretwo.com,
www.womensgolfunlimited.com)

TaylorMade-adidas
5545 Fermi Court
Carlsbad, CA 92009
(www.taylormadegolf.com)

Top-Flite Golf Company, Ben Hogan
425 Meadow St.
Chicopee, MA 01021
(www.benhogan.com)

TourShot (Bridgestone Sports)
1423 Lochridge Blvd., Suite G
Covington, GA 30012
(www.precept.com)

U.S. Kids Golf
3040 Northwoods Parkway
Norcross, GA 30071
(www.uskidsgolf.com)

Wilson Sporting Goods
8700 W. Bryn Mawr Ave
Chicago, IL 60631
(www.wilson.com)

Yonex
3520 Challenger Street
Torrance, CA 90503
(www.yonex.com)

Golf Club Shafts

Aldila
13450 Stowe Drive
Poway, CA 92064
(www.aldila.com)

Fujikura Composites
1493 Pointsettia Ave., Suite 139
Vista, CA 92081
(www.fujikuragolf.com)

Graphite Design
7919 St. Andrews Avenue
San Diego CA 92154
(www.gdintl.com)

UST
14950 FAA Blvd., Suite 200
Fort Worth, TX 76155
(www.ustgolfshaft.com)

Golf Club Grips

Golf Pride
16900 Aberdeen Rd
Laurinburg, NC 28353
(www.golfpride.com)

Karakal Golf Grips
945 Welsh View Dr., Suite C2
Newark, OH 43055
(www.karakalgolfgrips.com)

Tacki-mac Grips
3300 W. Desert Inn Rd.
Las Vegas, NV 89102
(www.tackimac.com)

Winn Grips
15648 Computer Lane
Huntington Beach, CA 92649
(www.winngrips.com)

Clubfitting

Accusport
4310 Enterprise Drive, Suite C
Winston-Salem, NC 27106
(www.accusport.com)

Callaway Golf
2180 Rutherford Road
Carlsbad, CA 92008
(www.callawaygolf.com)

Focaltron
830-A East Evelyn
Sunnyvale, CA 94086
(www.focaltron.com)

Henry-Griffitts
P.O. Box 1630
Hayden Lake, ID 83835
(www.henry-griffitts.com)

MacGregor Golf
1000 Pecan Grove Dr.
Albany, GA 31701
(www.macgregorgolf.com)

PING
P.O. Box 82000
Phoenix, AZ 85071
(www.pinggolf.com)

Golf Balls

Callaway Golf
2180 Rutherford Road
Carlsbad, CA 92008
(www.callawaygolf.com)

Maxfli
5545 Fermi Court
Carlsbad, CA 92009
(www.taylormadegolf.com)

Nike Golf
One Bowerman Drive
Beaverton, OR 97005
(www.nike.com/nikegolf)

Pinnacle
P.O. Box 965
Fairhaven, MA 02719
(www.pinnaclegolf.com)

Precept (Bridgestone Sports)
1423 Lochridge Blvd., Suite G
Covington, GA 30012
(www.precept.com)

Titleist
P.O. Box 965
Fairhaven, MA 02719
(www.titleist.com)

Top Flite
425 Meadow Street
Chicopee, MA 01021
(www.topflite.com)

Golf Shoes

adidas
5545 Fermi Court
Carlsbad, CA 92009
(www.taylormadegolf.com)

Bite Footwear
7120 185th Ave. NE
Redmond, WA 98052
(www.biteshoes.com)

Callaway Golf
2180 Rutherford Road
Carlsbad, CA 92008
(www.callawaygolf.com)

Champ
P.O. Box 735
289 Elm Street
Marlborough, MA 01752
(www.champsports.com)

Footjoy
P.O. Box 965
Fairhaven, MA 02719
(www.footjoy.com)

Nike Golf
One Bowerman Drive
Beaverton, OR 97005
(www.nike.com/nikegolf)

Oakley
One Icon
Foothill Ranch, CA 92610
(www.oakley.com)

Surefoot
1500 Kearns Blvd., Suite A-100
Park City, UT 84060
(www.surefoot.com)

Golf Bags

Izzo Golf
1635 Commons Pkwy.
Macedon, NY 14502
(www.izzo.com)

Mizuno
4925 Avalon Ridge Parkway
Norcross, GA 30071
(www.mizunousa.com)

Ogio
14926 Pony Express Road
Bluffdale, UT 84065
(www.ogio.com)

PING
P.O. Box 82000
Phoenix, AZ 85071
(www.pinggolf.com)

Sun Mountain
P.O. Box 7727
301 N First Street
Missoula, MT 59802
(www.sunmountain.com)

Golf Gloves

Etonic
260 Charles Street
Waltham, MA 02453
(www.etonic.com)

Footjoy
P.O. Box 965
Fairhaven, MA 02719
(www.footjoy.com)

Mizuno
4925 Avalon Ridge Parkway
Norcross, GA 30071
(www.mizunousa.com)

Nike Golf
One Bowerman Drive
Beaverton, OR 97005
(www.nike.com/nikegolf)

Titleist
P.O. Box 965
Fairhaven, MA 02719
(www.titleist.com)

Golf Gadgets

Brookstone
(www.brookstone.com)

Brush-T
8460 Higuera Street
Culver City, CA 90232
(www.brusht.com)

Divix Golf
7734 Arjons Drive
San Diego, CA 92126
(www.divixgolf.com)

Technasonic
3700 W. Morse Ave.
Lincolnwood, IL 60712
(www.technasonic.com)

Twilight Tracer, Sun Products
P.O. Box 398165
Edina, MN 55435
(www.twilighttracer.com)

Section 2: The Golf Course

Golf Courses

Augusta National Golf Club
P.O. Box 2086
Augusta, GA 30904

Henebry Photography
P.O. Box 21
La Quinta, CA 92253
(www.henebryphotography.com)

Highland Park
3300 Highland Ave. South
Birmingham, AL 35205

Merion Golf Club
450 Ardmore Ave.
Ardmore, PA 19003
(www.meriongolfclub.com)

Old Colliers Reserve
790 Main House Dr.
Naples, FL 34110

Olympia Fields
2800 Country Club Dr
Olympia Fields, IL
(www.olympiafieldscc.com)

Pebble Beach Golf Links
P.O. Box 1767
Pebble Beach, CA 93953
(www.pebblebeach.com)

The Riviera Golf Club
1250 Capri Dr.
Pacific Palisades, CA 90272
(www.theriveriagolf.com)

Rustic Canyon
15100 Happy Camp Canyon Rd.
Moorpark, CA 93021

Windsong Farm
18 Golf Walk
Independence, MN 55359

Golf Course Design and Maintenance

ArborCom Technologies
13435 South McCall Road, Suite 230
Port Charlotte, FL 33981
(www.arborcom.ca)

(Robert) Cupp Design
5457 Roswell Rd., Suite 103
Atlanta, GA 30342
(www.cuppdesign.com)

Golf Course Superintendents Association of America (GCSAA)
1421 Research Park Drive
Lawrence, KS 66049
(www.gcsaa.org)

Hancor
401 Olive Street
Findlay, OH 45840
(www.hancor.com)

John Fought Golf Course Architecture
5010 Shea Blvd., Suite A-17
Scottsdale, AZ 85254
(www.foughtdesign.com)

Nicklaus Design
11780 U.S. Highway One, Suite 500
North Palm Beach, FL 33408
(www.nicklaus.com)

Signature Control Systems
4 Mason, Suite B
Irvine, CA 92618
(www.signaturecontrolsystems.com)

United States Golf Association (USGA)
PO Box 708
Far Hills, NJ 07931
(www.usga.com)

Golf Cars

Club Car
P.O. Box 204658
Augusta, GA 30917
(www.clubcar.com)

E-Z-Go
P.O. Box 388
Augusta, GA 30903
(www.ezgo.com)

Yamaha
1000 Highway 34 East
Newnan, GA 30265
(www.yamaha.com)

GPS and Range Finders

Bushnell
9200 Cody
Overland Park, KS 66214-1734
(www.bushnell.com)

Nikon
1300 Walt Whitman Rd.
Melville, NY 11747
(www.nikon.com)

ParView
1856 Apex Rd.
Sarasota, FL 34240
(www.parview.com)

ProLink
7970 S. Kyrene Rd.
Tempe, AZ 85284
(www.goprolink.com)

ProShot
13865-A Alton Parkway
Irvine, CA 92618
(www.proshotgolf.com)

SkyGolf
SkyHawke Technologies, LLC
P.O. Box 2960
Ridgeland, MS 39158
(www.skygolfeps.com)

UpLink
9508 Jollyville Rd., Suite 200
Austin, TX 78759
(www.uplinkgolf.com)

Section 3: The Golfer

Golf Training

Astar
11722-D Sorrento Valley Road
San Diego, CA 92121
(www.astarls.com)

CoolGrip
4611 Teller Ave.
Newport Beach, CA 92660
(www.coolgripgolf.com)

David Leadbetter Golf Academies
DLGA Headquarters
5500 34th Street West
Bradenton, FL 34210
(www.leadbetter.com)

Debbie Steinbach's Venus Golf
78-110 Calle Norte
La Quinta, CA 92253
(www.venusgolf.com)

Electronic Arts (EA Sports)
209 Redwood Shores Pkwy
Redwood City, CA 94065
(www.ea.com)

Golf Around the World
1396 N. Killian Drive
Lake Park, FL 33403
(www.golfaroundtheworld.com)

Jim McLean Golf School
(www.jimmclean.com)

Medicus Dual 2000
(www.medicus2.com)

Karen Palacios-Jansen
P.O. Box 3354
Mooresville, NC 28117
(www.swingblade.com)

Pelz Golf
1310 R.R. 620 South, Suite B-1
Austin, TX 78734
(www.pelzgolf.com)

Power Wrist
P.O. Box 806115
St. Clair Shores, MI 48080
(www.powerwrist.com)

The Putting Arc
(www.theputtingarc.com)

Strolf
875 Eglinton Ave. West
Toronto
(www.strolf.com)

Swing Solutions
1166 Triton Drive, Suite 200
Foster City, CA 94404
(www.swingsolutions.com)

Tac-Tic
14841 Curtis Circle
Sonora, CA 95370
(www.tac-tic.com)

United Turf Industries, SofTrak Putting Greens
(www.unitedturf.com)

V1
38777 West Six Mile Rd., Suite 311
Livonia, MI 48152
(www.v1golf.com)

The Whippy TempoMaster
10545 Lennox Lane
Dallas, TX 75229
(www.tempomaster.com)

Golf Apparel

adidas
5545 Fermi Court
Carlsbad, CA 92009
(www.taylormadegolf.com)

Ashworth
2765 Loker Ave. W.
Carlsbad, CA 92008
(www.ashworthinc.com)

Great White Shark Enterprises, Greg Norman
Collection
501 N A1A
Jupiter, FL 33477
(www.shark.com)

Nike Golf
One Bowerman Drive
Beaverton, OR 97005
(www.nike.com/nikegolf)

Polo
650 Madison Ave.
New York, NY 10022
(www.polo.com)

Sunderland of Scotland
20844 Plummer St.
Chatsworth, CA 91331
(www.sunderlandgolf.com)

Titleist
P.O. Box 965
Fairhaven, MA 02719
(www.titleist.com)

Zero Restriction
80 South Prospect St.
Hallam, PA 17406
(www.zrgolf.com)

Sunglasses

adidas
5545 Fermi Court
Carlsbad, CA 92009
(www.taylormadegolf.com)

Bolle
9200 Cody
Overland Park, KS 66214
(www.bolle.com)

Maui Jim
721 Wainee St.
Lahaina, HI 96761
(www.mauijim.com)

Nike Golf
One Bowerman Drive
Beaverton, OR 97005
(www.nike.com/nikegolf)

Oakley
1 Icon
Foothill Ranch, CA 92610
(www.oakley.com)

Sun Lotions

Proderma
P.O. Box 79
Spring Lake, NJ 07762
(www.prodermaproducts.com)

Fitness Training

Back to Golf
7075 N. Howard St. #105
Fresno, CA 93720
(www.backtogolf.com)

Golf Health and Performance Center, Rob Mottram
5690 Cancha De Golf
Rancho Santa Fe, CA 92091

Momentus Golf
111 North Locust
Winfield, IA 52659
(www.momentusgolf.com)

Swing Wave
15222 King Road, Suite 403
Frisco, TX 75034
(www.swingwavegolf.com)

The Mental Game

Coaching for the Future
2712 East Mountain View Road
Phoenix, Arizona 85028
(www.coachingforthefuture.com)

Golf is Not a Game of Perfect, by Bob Rotella (Simon and Schuster, 1995)

Dr. Deborah Graham/Jon Stabler: Golfpsych
Sportpsych, Inc.
P.O. Box 1976
Boeme, TX 78006
(www.golfpsych.com)

HeartMath
14700 West Park Ave
Boulder Creek, CA 95006
(www.heartmath.com)

Other Resources

Golf Digest
(www.golfdigest.com)

Golf for Women
(www.golfdigest.com/gfw)

Golf Magazine
(www.golfonline.com)

Golf Illustrated
(www.golfillustrated.com)

The Golfer Magazine
551 Fifth Ave.
New York, NY 10176

Golf Tips
(www.golftipsmag.com)

Golfweek
(www.golfweek.com)

Links Magazine
(www.linksmagazine.com)

The PGA of America
100 Avenue of the Champions
Palm Beach Gardens, FL 33418
(www.pga.com)

(561) 624 8400

Peterson's Golfing

Sports Illustrated Golf Plus
(www.cnnsi.com/golf)

Travel and Leisure Golf
(www.travelandleisure.com/tlgolf)

Index